MARINE LIFE
OF THE CARIBBEAN SEA

MARINE LIFE
OF THE CARIBBEAN SEA

LAWSON WOOD

Princeton University Press
Princeton and Oxford

Published in the United States, Canada, and the Philippines in 2026
by Princeton University Press
41 William Street, Princeton, New Jersey 08540
press.princeton.edu

First published as *Marine Life of the Caribbean* in the United Kingdom
by Bloomsbury Publishing Plc in 2026

Library of Congress Control Number: 2026933523
ISBN: 978-0-691-29245-8
ISBN (PDF): 978-0-691-29425-4

Designed by Austin Taylor
Map by Julian Baker
Cover photographs © Lawson Wood

Printed and bound in China by Toppan Leefung Printing

1 3 5 7 9 10 8 6 4 2

MIX
Paper | Supporting responsible forestry
FSC® C104723

CONTENTS

THE CARIBBEAN 7

FISH 30

OTHER VERTEBRATES 114

INVERTEBRATES 120

PLANTS AND ALGAE 208

UNITED STATES OF AMERICA
Bermuda
N
S
W
E
ATLANTIC OCEAN
FLORIDA
Grand Bahama
Abaco Islands
Gulf of Mexico
Miami
Eleuthera
Bahamas Islands
Andros
Caicos Islands
Turks Islands
MEXICO
Cozumel
CUBA
Chinchorro Bank
Cayman Islands
PUERTO RICO
Leeward Islands
Roatan
JAMAICA
HAITI
DOMINICAN REPUBLIC
Windward Islands
HONDURAS
CARIBBEAN SEA
Bonaire
Curacao
Los Roques
Aruba
NICARAGUA
Santa Marta
San Andres
COSTA RICA
PANAMA
VENEZUELA
Virgin Islands
Anguilla
St. Martin/St.Maarten
St. Barthelemy
Saba
Barbuda
St. Croix
Antigua
St. Eustatius
Montserrat
St. Kitts
Nevis
Guadeloupe
Dominica
Martinique
St. Lucia
St. Vincent
The Grenadines
Barbados
Grenada
Tobago
Trinidad
VENEZUELA
COLOMBIA
PACIFIC OCEAN
ECUADOR
PERU
BRAZIL
0
1000 km
0
500 miles
International boundaries
Reef

THE CARIBBEAN

I FIRST VISITED FLORIDA IN 1982, and then returned for a longer trip through Florida, Mexico and the Cayman Islands in 1991. However, my first real introduction to the complexity of marine life in the Caribbean Sea came after a request to write a diving guide to the Cayman Islands. This successful project led me to tackle a number of other books covering the whole of the Caribbean. My wife and I were married in Bermuda and made Cayman Brac our home for almost ten years, allowing us to explore and dive a variety of the 7,000 islands that make up the 13 countries in the Caribbean. The species included in this volume are those likely to be encountered by the visitor. All are found in fairly shallow water, averaging less than 18m (60ft).

The region's English name derives from the indigenous Carib people who first explored and settled amidst this vast archipelago, which covers more than 2.7 million km² (1.06 million square miles). The Arawak and Lucayan people had previously explored some islands, leaving telltale carvings on ancient coral rock. These travellers originated from Central and South America, around 6,000 years ago. In the northern Gulf of Mexico during the Pleistocene period and during the last Ice Age, the sea level was more than 100m (330ft) lower, making Florida more than twice the size it is now. Today, we can explore the ancient caves and caverns which were formed in this period, including those in the Cenoté, in Mexico's Yucatán Peninsula, the massive Blue Hole in Belize's outer atolls, and of course the equally impressive blue holes in the Bahamas. Local Floridian indigenous groups also influenced the northern Caribbean islands, but most scientists agree that the greatest influence of migration came from South America in two distinct waves; successive DNA studies have proven this.

The Caribbean is much smaller than many people understand, as it is bordered to the north by the Greater Antilles, including the islands of Cuba, Haiti/Dominican Republic, and Puerto Rico. Located entirely within the Caribbean Sea are the island countries of Jamaica and the Cayman Islands. To the west are the Central American countries of Mexico, Belize,

▶ Little Cayman Island in the central Caribbean.

Guatemala, Honduras, Nicaragua, Costa Rica and Panama. To the south, in mainland South America, are found Colombia and Venezuela. Off the northern coast of Venezuela are the Dutch Antilles of Aruba, Curaçao and Bonaire. In the south-east corner is the island nation of Trinidad and Tobago and north-east of this is the fairly isolated island of Barbados.

Moving north and north-west, we reach the eastern group of Caribbean islands, the Lesser Antilles, which is made up of the Windward and Leeward Islands. The Windward Islands are Grenada, St. Vincent and the Grenadines, St. Lucia, Martinique and Dominica. In the northern arc are the Leeward Islands, comprising Guadeloupe, Montserrat, Antigua, Barbuda, St. Kitts, Nevis, St. Eustatius, St. Barthélemy, Saba, Anguilla, St. Martin/Sint Maarten, the British Virgin Islands and the US Virgin Islands.

The western North Atlantic Ocean is located above the Antilles and encompasses the Turks and Caicos, and the Bahamas. These countries are bathed in warm water from the Caribbean which flows constantly through the Gulf of Mexico and up the eastern seaboard of the United States of America. The influence of the Gulf Stream is such that it has contributed to the most northerly coral reefs in the world, found around the shores of Bermuda. The Turks and Caicos, the Bahamas and Florida are often regarded as being part of the Caribbean.

The Atlantic Ocean is by far the greatest influence on the Caribbean Sea, but there is also a threat of influence from the Pacific Ocean through the construction and later the widening of the Panama Canal (see box on page 10). Now, Caribbean and Atlantic species of marine life are being recorded in Gatun Lake, including West Indian Manatees, which have been sighted towards the Pacific side of the canal.

Through the various ice ages and continental upheaval, the Caribbean now has the most northerly coral atolls on the planet, with Arrecife Chinchorro off Mexico's

Blue holes

Blue holes are found in shallow limestone carbonate platforms, predominantly in the Bahamas, Florida, Belize and the Yucatán Peninsula in Mexico. They were formed during the last Ice Age when the sea level was as much as 100m (330ft) lower than it is now. Whilst dry, they were subject to erosion, creating great caverns of stalactites and stalagmites, which gradually filled back up again as the sea level rose once more.

Roughly circular in shape, blue holes have near vertical walls and are usually very deep, resulting in a colour change of water when viewed from above, with pale colours in the shallows and deeper indigo blue in the centre. Two of the deepest blue holes in the Caribbean are Dean's Blue Hole to the west of Clarence Town in Long Island, Bahamas, and the Great Blue Hole in Belize which is more than 300m (1,000ft) in diameter. In 2003, another huge blue hole, technically a sink hole, was discovered in Chetumal Bay. Explorations in 2023 revealed that this hole, called Taam Ja' Blue Hole, is around 274m (900ft) deep and spans an area of 13,660m² (147,000ft²).

◂ The Great Blue Hole in Belize is one of the world's natural wonders.

▲ A dive boat above a tropical reef festooned with sponges, corals and sea fans.

Yucatán Peninsula regarded as being the largest. Other atolls are found off the coast of Belize and north of Venezuela. Bermuda should also be included as it has formed, over some 30 million years, at the tip of an extinct volcano; although it is actually classed as a pseudo-atoll, it has the same characteristics as true atolls. During this process, as the volcano reached upwards, sands collected and were then compressed and turned into rock by a process known as lithification.

The Belize/Mexico Barrier Reef is the second-largest barrier reef in the world and the largest in the northern hemisphere. The second-largest barrier reef in the Caribbean (and third largest in the world) extends from Cuba to the British Virgins island of Anegada.

Undoubtedly the Caribbean's marine habitats and the creatures and plants which live on, in and around them are greatly influenced by the might of the Atlantic Ocean. Many of the hundreds of species of Caribbean fish, invertebrates and algae are found nowhere else in the world, but we can also find creatures which are recorded in the Pacific Ocean and even as far away as the Indian Ocean and Mediterranean Sea.

The biodiversity of the region has also been strongly influenced by the rise of pollution, the environmental impact of the chemical industry, intensive agriculture, aquaculture, oil and gas platforms, coastal erosion, coastal development, and large-scale reclamation of the sea (to the detriment of the coastal mangrove forests which once protected the islands from the mighty force of the fairly predictable annual hurricanes). Climate change also impacts the marine ecosystems, with more storms out of season, and other threats include the arrival of invasive species such as Lionfish, which is wreaking havoc throughout the entire region.

Nevertheless, the diversity of Caribbean marine life is vast and features a staggering number of invertebrates and fish, as well as turtles, dolphins and other cetaceans, and seabirds in their countless thousands. Large land-based seabird colonies are indicative of the profusion of fish to be found in certain areas, and sizeable colonies of boobies, pelicans, flamingos, terns and gulls can be found on most Caribbean islands. Some of the world's largest cetaceans also make the Caribbean their home.

Tides and currents

Analysis of the Caribbean Sea's tides and currents has been carried out by NOAA (National Oceanic and Atmospheric Administration), using material collected from 45 different locations throughout the zone. These results show that the Caribbean Sea has a microtidal range of less than 20cm (8in) with the main influence coming from magnetic variations from the moon's phases. The greater influence comes from the North Equatorial Current, which sweeps west across the Atlantic from Africa and enters the Caribbean through the Windward and Leeward Islands chain. The current splits in two here, with the northern arm continuing up through the Bahamas and north up the eastern United States. The southern arm of the current passes through the central Caribbean where it warms up and gathers speed as it passes up through the narrow passage between Cuba and the Yucatán Peninsula, and morphs into the Gulf Stream. From here it channels through the Gulf of Mexico, around the southern tip of Florida, and moves through the Turks and Caicos and the Bahamas. It then rejoins the northern arm of the Atlantic current and continues on to Bermuda and east across the Atlantic, as the North Atlantic Gyre.

It was these consistent currents which brought Christopher Columbus to the West Indies, and aided the Spanish treasure fleets in their return to Europe.

The temperature of the Caribbean Sea is typically warm, being fed by the heated North Equatorial Current. The average water temperature is around 27°C (80°F) and varies by as little as 3°C (5.4°F) throughout the year. Northern latitudes do get colder due to colder air passing south from North America, but in general the temperature is fairly consistent.

Panama Canal

Ever since the Spanish and subsequent piratical nations traversed the Panama Isthmus with their ill-gotten gains, there had always been dreams of connecting the various lakes and waterways in southern Central America to join the Pacific to the Caribbean. Following the success of the Suez Canal, the French started construction of the Panama Canal in the 1880s. In nine years, the French attempt went bankrupt and saw the loss of 20,000 workers, mainly due to malaria, yellow fever and other tropical diseases. After Panama gained its independence from Colombia in 1903, the American government negotiated a deal with the new Panamanian government to complete the canal, and the first ship made the journey from ocean to ocean in 1914. Although the difference in sea level is minimal at only 20cm (8in) between the two oceans, locks are required to lift and lower the ships to pass through Gatun Lake, a key part of the canal that sits at a higher elevation. At 82 km (51 miles) long, the Panama Canal takes around 8–10 hours to traverse, saving approximately five months and 12,875km (8,000 miles) of sea travel on a journey from New York to California. The canal also presents a potential threat to the marine life of the Caribbean, with the possibility of the transfer of marine species between the Caribbean and Pacific oceans.

THE HABITATS

The various habitats are as diverse as they are in any specialised environment, and although there are some very deep water sections and many commercial fishing grounds, we shall only concern ourselves in this volume with the shallower depths that can be explored by scuba diving, snorkelling or rockpooling.

Rocky shores

These offer a firm substrate base for hardy algae and other more sedentary forms of life to attach to. The particular selection of marine life will depend on the type of rock. Some softer rocks can be bored into, whilst other glaciated granite rocks only have a slight scattering or patina of animal life as there is nowhere for them to get a firm grip. Exposed locations will always be strongly affected by prevailing storms and adverse weather conditions, and this factor will also determine the types of marine life to be found. Chitons, urchins, limpets and barnacles are perhaps the most common rocky shore inhabitants. Much of the shoreline is made up of fossilised limestone, known as 'ironshore' as this incredibly hard substrate will 'ring' like metal if struck by another rock. Inshore from here are either sandy beaches or coral rock debris that has been deposited by hurricanes, as it smashes the inshore coral reefs.

Sandy shores

Here, low tide provides the first glimpse of the underwater world, with washed up small shells, seaweed and coral debris. Birds forage along the shore and some fishermen cast their lines into the shallow bays. This habitat is incredibly volatile as it is subjected to severe pounding by tropical storms and the more powerful hurricanes. Subsequently, only the hardiest of animals can live in this zone, but thankfully, most sandy bays are fringed by rocky areas and rockpools. Wind-blown sand has created huge sand dunes and even islands made only of sand particles.

▼ A typical ironshore environment with very hard rock creating small rockpools.

Rockpools

These pools of seawater are captured in rocky crevices after the tide recedes. They are directly on the shoreline, within what is known as the littoral zone. Mainly found on ironshore, these mini-oases of life are akin to aquaria, but the creatures living within them are incredibly hardy, being able to withstand both a rapid change of salinity (after heavy rainfall in a shallow pool) and changes in water temperature (especially at the height of summer, when the water temperature in shallow pools can rise alarmingly over 6–8 hours before the advancing tide covers them again). Rockpools will usually have several types of anemone, small blennies and tiny crabs, shrimps and nudibranchs, with the rocky exposed sides being grazed upon by snails.

Rocky cliffs

Submarine cliffs are well known for their colourful fields of sponges, and underhanging ledges bristling with starfish. They are home to coral crabs, coral sea fans that stretch out into the current, black corals, wire corals, brilliant red finger sponges, and huge barrel sponges. If you take a closer look, you will find squat lobsters, gobies, blennies and various nudibranchs. These vertical walls are patrolled by various jacks, turtles and even reef sharks. Scuba divers will often find caves and caverns associated with these cliff faces; these features, combined with a deep vertical wall, make for a very enjoyable dive.

Caves and caverns

Wherever the coastline is made up of a mixture of ancient limestone and softer rock types, and where geological upheaval has shifted the strata, caves and caverns are found. Cayman Brac, part of the Cayman Islands, is often likened to a huge slice of Swiss cheese as it is wedge-shaped and riddled with caves and caverns. Underwater, these caverns are always interesting, with good populations of nocturnal animals.

Shipwrecks

Shipwrecks are very important habitats, especially when found in areas that would otherwise be rather poor for marine life. Many have been sunk deliberately as 'artificial reefs' – made safe for scuba divers' explorations. Wrecks provide an important holdfast for soft and hard corals, encrusting algae, anemones, sponges, sea squirts and hydroids, and are home to a number of commercially important species of fish which are a natural draw for both fishermen and scuba divers.

◂ Shallow caves fill with Silverside Minnows in the summer months.

Thankfully most are located within Marine Protected Areas (MPAs), and are left intact for the enjoyment of everyone.

Oil and gas rig support legs

In the northern regions of the Gulf of Mexico, there is a marine sanctuary known as the Flower Garden Banks, which protects a range of coral reefs. Divers can access the sanctuary from various ports in the southern United States. Just beyond its boundaries are numerous oil and gas rigs, and their steel supporting legs, lattice structures, pipelines and mooring platforms have become colonised by diverse marine life over the years. In some instances, the marine growth is so great that these legs have to be cleaned periodically. However, they give an almost perfect zonation profile with barnacles and snails in the shallows, ranging down through various algae to sponges, soft corals, hard corals, sea fans and anemones.

Piers and bridges

The most obvious of the built structures which abut the seashore, piers, bridges and breakwaters are often the only shore access on significant stretches of coastline, as many islands have virtually no access for shore diving. The structures have evolved over the years as boat traffic has changed in shape, size and frequency of visits, with cruise ships being an increasingly frequent presence. Many of these sites are only accessible at night as port authorities ban diving during the day when shipping is using the piers. Piers such as Frederiksted Pier on St. Croix, the Cruise ship pier in Cozumel, the Town Pier and Salt Pier in Bonaire, the Port Pier on Cayman Brac, and Blue Heron Bridge in Florida are all superb.

▲ Ships sunk as artificial reefs are a huge draw for marine life and divers.

▼ The Salt Pier in Bonaire is a favourite with divers.

Sand

You will find the tracks of marine snails and other creatures on sandy seabeds, and many burrowing animals make this region their home, including starfish, heart urchins, sand dollars, shells, burrowing anemones and fan worms, eels and many types of fish, particularly garden eels, razorfish and Flying Gurnard. Eagle Rays often hunt in these areas.

Mud

Deeper seabeds and those affected by river run-off build up muddy layers, as detritus is washed into them during periodic winter storms and heavy rainfall. Burrowing animals, such as various molluscs, mud crabs, brittle starfish, burrowing anemones, fan worms, flatfish and mantis shrimps, are typical of this habitat and are all excellent photographic subjects.

Coastal wetlands

As nurseries for marine fish, coastal wetlands are extremely productive ecosystems. Their waters trap carbon and fix substantial amounts of nitrogen annually, a process that converts nitrogen into compounds such as ammonia that can then be used by plants. This forms the basis for wetland and marine food webs, which include decomposing organic matter, plants, crustaceans, worms, fish, birds and many other organisms.

Mangrove forests

Mangrove forests constitute a very important marine habitat in the Caribbean. These sturdy trees were historically found around almost every shoreline of every island in the Caribbean. Hurricanes take their toll, but by far the greatest losses have been through human intervention, primarily clearing mangroves to develop land for leisure facilities, commercial enterprises and housing, mainly for the tourism industry.

Mangroves are vital in many different ways, but principally they are the Caribbean's greatest nursery for dozens of species of fish, sharks and commercially valuable invertebrates such as crabs and spiny lobsters. They quell water motion and reduce wave-induced coastal erosion.

Tidal rapids

The Caribbean and wider Gulf zone are largely influenced by the North Atlantic Gyre (moving at approximately 2.5km/h or 1.5mph) which sweeps warm water northwards and eastwards to bathe the shores of Bermuda. The Gulf Stream, part of this system, travels fastest on the surface at around 9km/h (5.6mph) and has an average speed of 6.4km/h (4mph). It transports more water than all of the world's rivers combined.

Common in a number of localities between various islands, tidal rapids are home to algae, sponges, tube worms, anemones, sea fans, soft and hard corals and other invertebrates which feed on the swiftly moving plankton passing on their way. Strong currents are to be expected on every dive off the west coast of Cozumel where the Gulf Stream is funnelled through the gap between the island and the Yucatán mainland. There are a number of drift dives through the Columbus Passage between Trinidad and Tobago which are quite exhilarating, as well as several others around the north-eastern tip of Tobago.

The water column and plankton

Plankton (both phytoplankton and zooplankton) provides a fundamental role in the food chain of pelagic (oceanic) wildlife. The abundance of plankton is strongly influenced by factors such as depth, tidal mixing and temperature stratification, all of which determine the vertical stability of the water column. Plankton blooms begin well offshore in March, when nutrient levels are high, and the seawater warms even more as it crosses the Atlantic. Within the Caribbean area, blooms are dominated by diatoms, which spread westwards throughout the Caribbean by April. Plankton has a fundamental role in the food chain and is the main diet of many species of benthic and pelagic wildlife, including jellyfish and cetaceans.

DISTRIBUTION AND FEATURES OF CARIBBEAN REEFS

All of the fish, invertebrates, algae, plants and mammals in this book are associated with the complexity of the Caribbean reef system. Coral reefs are found in every portion of the Caribbean and Gulf of Mexico, as far north-east as Bermuda, south to isolated Barbados in the outer margins of the Caribbean basin, and around all of the surrounding shores from Venezuela in the south to the United States mainland in the north.

TYPES OF CORAL REEFS

Reefs are classified according to their shape, position and composition of corals.

Patch

Patch reefs form on old rocky or fossilised coral outcrops, usually in waters less than 6m (20ft) deep. Large hard corals such as brain and boulder corals predominate on these reefs and grow from the calcium carbonate skeleton which they secrete. Small sea fans are also found in these regions and patch reefs are very important for protecting the fragile coastline from tropical storms. The United States Virgin Islands have superb patch reefs all around all their islands. Patch reefs are often found within a protected lagoon or atoll.

Fringing

Fringing reefs are, as the name suggests, attached to the shoreline and 'fringe' an island with a protective coral reef. Usually with a very shallow top, they have a mini wall that descends to about 12m (39ft) and thence over a protected sand patch where garden eels and Flying Gurnards can be found. Some such reefs then adjoin a barrier reef and plunge into the depths, such as Jackson's Bight to the north of Little Cayman Island.

▼ Snorkelling the inner protected reef at Point of Sand on Little Cayman Island.

▲ Spur and groove reefs are a predominant feature around many Caribbean islands.

Barrier

As the name suggests, these important reefs form a natural, living barrier which runs parallel to the shore and therefore protects the shoreline from the worst of the tropical storms. The second-largest barrier reef in the world, the Mesoamerican Barrier Reef, is found in the far west of the Caribbean. It stretches across more than 900km (560 miles) from Honduras to Isla Contoy in Mexico.

Bank

Bank reefs actually include several other reef formations such as reef flat, spur and groove, and fore reef. Bank reefs are larger than patch reefs, although in essence they are much the same, having similar corals and therefore similar characteristics and a similar range of species living in, on and around them.

The **reef flat** is the area closest to shore and behind the bank reef. It is usually made up of sand patches and broken coral skeletons and is regarded as quite inhospitable due to its turbulent nature and sometimes dangerous sea conditions.

Spur and groove reefs are best seen from the air. They are formed from wave-led power surges, which create 'fingers' or spurs of coral and grooves of sand and coral rubble that concentrate the tidal surges. The spurs are topped with small sea fans, soft corals and many different sponges, making homes for blennies, gobies, nudibranchs and many different species of wrasse and grouper. The grooves are generated by wave action and inevitably have broken coral debris, gravel and sand littered along their lengths. They are populated with jawfish, wrasse and even Nurse Sharks.

The **fore reef** is the outer edge of the reef system, which can take the full brunt of any storm. It is from here that the reef structure will plunge to deep water, such as off the Cayman Islands and US Virgin Islands. The reef edge is predominantly made of Boulder Star Coral, but as depth limits the range of species, more plate-like corals are found. These increase the surface area available for the corals' symbiotic, photosynthesising algae.

DIVING THE CORAL REEFS OF THE CARIBBEAN

Coral reefs are the single most important resource to be found in the Caribbean. There are literally thousands of reefs throughout the entire region, and by all accounts only around 20 per cent of them are ever visited by divers. This section describes the reefs that are most popular with visiting scuba divers and those which, quite rightly, have been declared as national parks or marine sanctuaries of some kind or other. Most are fringing reefs, in that they are directly attached to the shore.

Bermuda

To the far north-east of the American mainland is Bermuda. It lies 1,040km (650 miles) from Cape Hatteras (the most easterly point of the United States) and 6,000km (3,750 miles) from mainland Europe, and holds the most northerly coral reefs in the world. The main 'boiler' reefs are actually formed from calcareous algae but are studded with soft and hard corals and sea fans.

USA, Florida and northern Gulf of Mexico

Whilst many people associate the Florida reefs with the greater Caribbean, they are actually an arc of chain and patch reefs that run parallel to the eastern shores of the Florida Keys. As you travel north along the coast, the proportion of patch and bank formations increases. One of the best shore dives in Florida is Blue Heron Bridge in North Lauderdale, which (despite a lack of corals) has an amazing diversity of marine life. As you travel south along the Keys, corals and sponges cover a number of shipwrecks that are available for sports divers. In Key Largo, the John Pennekamp Coral Reef State Park is the United States' first marine conservation area. Created in 1963, it covers around 178 square nautical miles of corals, seagrass beds and mangrove forests. The park was added to the National Register of Historic Places on 14 April, 1972.

In the northern Gulf of Mexico, the Flower Garden Banks south of Galveston, Texas, are located 100 nautical miles offshore and are the most northerly coral reefs in the United States. This series of bank reefs is home to around 300 species of fish, and 20 coral species have been recorded.

▸ Sea fans at the edge of the drop-off often have crinoids on them, also filter feeding in the gentle current.

Central America

Continuing in an anti-clockwise direction, we arrive at Isla Contoy just north of Cancún in the Caribbean Yucatán. This is where the Mesoamerican Barrier Reef terminates. Divers can enjoy the delights of Whale Sharks and Sailfish just offshore, but it is the long patch reefs between Cancún and Isla Mujeres that are the most popular with divers and snorkellers.

Cozumel Island has an enviable reputation for drift diving from the south to the north along the west coast of the island. Fed by the constant movement of the Caribbean Current, the reefs have healthy corals, but it is the sponges that are most dramatic.

The offshore reefs of Belize are world class, and Turneffe Reef, Glover's Reef and Lighthouse Reef, with the amazing Great Blue Hole, appear to have been custom-made for diving, with some exceptional coral reefs and a huge selection of all the marine life found in the Caribbean.

Further to the south is the beginning of the Mesoamerican Barrier Reef around Cayos Cochinos, or the Hog Islands, in Honduras. The islands to the north of the mainland are Utila, Roatan and Guanaja, and are collectively known as the Bay Islands. I first dived these reefs back in 1998 with the Aggressor Fleet International. This offshore bank reef at Little Coco and Marjorie's Bank is fairly deep, starting at around 18m (60ft), but is home to an amazing array of pelagic fish and deeper-water corals.

The reefs off the north shore of Colombia have been stunted somewhat due to the freshwater run-off from the mainland, but there are a couple of popular islands which have good-quality fringing reefs.

Off the north coast of Venezuela are the three Dutch Antilles islands of Aruba, Bonaire and Curaçao (the ABCs). Bonaire tends to get the lion's share of divers as the fringing reef is very popular for shore diving. Offshore to Klein Bonaire are huge fields of Staghorn Coral, and this is one of the few places where I regularly find frogfish. There is a similar fringing reef to the south of Curaçao, but access is difficult from the shore and these dives have to be done by boat.

Along the north shore of Venezuela lies a massive stretch of offshore patch reefs which, if they weren't so far from the shore, might be classified as a barrier reef. There is also an atoll-like structure called Los Roches 125km (80 miles) offshore with more than 50 islands and literally hundreds of small coral reefs, which are sometimes dived from Curaçao.

▼ Although quite barren on land, the island of Bonaire has superb reefs and crystal-clear waters popular with shore divers.

▲ Giant Brain Coral in Tobago.

Windward Islands

The southeasternmost islands in the Caribbean are known collectively as the Windward Islands and are comprised of St. Lucia, Grenada, St. Vincent and the Grenadines (Bequia, Mustique, Canoun, Mayreau, Tobago Cays, Union, Carriacao and Ronde), Martinique and Dominica. Some people also include Trinidad, Tobago and Barbados in the Windward Islands group. All of these islands have very diverse reef structures and all are home to species more influenced by the western Atlantic and South America. However, they also host the array of life found in the rest of the Caribbean. Tobago in particular is directly in the Orinoco current. By the time the current reaches the Speyside region of the island, the waters have cleared, but they are still full of all the nutrients that have created the largest brain corals on the planet.

Leeward Islands

The Leeward Islands continue the arc of the volcanic islands running north to north-west of the eastern Caribbean. This chain comprises Antigua, Barbuda, St. Kitts, Nevis, Sint Maarten/St. Martin, St. Eustatius, St. Barthélémy, Saba, Anguilla, Montserrat and Guadeloupe. These islands are influenced by the Western Atlantic Gyre, but are in a more sheltered position and the greater Caribbean pushes its massive diversity of life all over their coral reefs. The islands sit atop a vast underwater volcanic mountain ridge, which was completely above sea level at one time. It should be noted that this volcanic chain is still active and, in the late 1990s, Montserrat's Soufrière Hills volcano threw at least 10 million cubic metres of volcanic material over 3km (2 miles) high, devastating the island.

The northern Caribbean

West of the Leeward Islands are the British Virgin Islands, comprising Anegada, Tortola and Virgin Gorda. South of the Sir Francis Drake Channel are the islands of Ginger, Cooper, Salt (where the wreck of the RMS *Rhone* is located), Norman and Peter Island. North and west of Tortola is another massively diverse group of islands including The Dogs (which are named after the noise created by the now extinct Caribbean Monk Seals).

The massive island of Puerto Rico lies 95km (60 miles) west of the Virgin Islands and around 1,800km (1,100 miles) south-east of Miami. Off its eastern coast are the popular diving islands of Culebra and Vieques, which have excellent fringing reefs. The north coast of Puerto Rico has limited reef structures, but the south and east are

▲ Great Hammerhead Sharks come close to shore in Bimini in the Bahamas.

the most popular. All of these are part of the United States, as are the US Virgin Islands with its three main islands of St. Thomas, St. John and St. Croix. St. Croix in particular has an amazing fringing reef as well as very well-developed patch reefs.

Continuing west is Cuba, the largest island in the Caribbean, which has extensive very good coral reefs all around its coast. The far west at Maria la Gordo is very similar to the west coast of Cozumel as it sits out into the Gulf Stream, and is home to amazing coral and sponge structures. To the south of the mainland is a region world-renowned for its beautiful reefs, which are relatively rarely visited by divers, yet are home to an amazing amount of marine life. The Gardens of the Queen (Jardines de la Reina) are exquisite and stand at the edge of the island platform with the outer (south-facing) wall plunging into the depths. The north shore of Cuba is in the Gulf of Mexico and the south is situated solely in the Caribbean.

The Caribbean islands

Jamaica is the largest island located solely within the confines of the Caribbean Sea. The northern shore in particular has an extensive fringing reef very similar to that found on Grand Cayman Island, with huge coral outcrops, surge channels, canyons and tunnels. There are similar reefs along the south shore, but the greater Caribbean Sea to the south of Jamaica has a staggering number of patch and bank reefs. Here are the hidden, shallow reefs where much of the Spanish treasure fleet fell foul.

The Cayman Islands have long been at the top of the list for scuba divers exploring the Caribbean. There are three islands: Grand Cayman, Little Cayman and Cayman Brac. There is also an underwater fourth island called Twelve Mile Bank, which is an ancient bank reef atop an even more ancient volcanic outcrop. I have dived here onto the remains of a sunken oil barge. This bank is home to huge barrel sponges and large extensive coral heads. All three of the main islands have excellent spur and groove reefs, as well as a barrier reef and patch reefs, and, of course, the famous Cayman Walls which lie to the north of all three islands. Corals here are in excellent condition and all are home to a mind-boggling amount of marine life.

The Gulf Stream islands

The Bahamas are often referred to as the Islands in the Stream and indeed the most westerly island of Bimini is exactly that, as it sits directly in the path of the northwards-pushing Gulf Stream. This group consists of around 3,000 islands and coral cays, which all sit atop a massive limestone platform that was once well above the current sea level. Sitting astride the Tropic of Cancer, every type of coral reef in the Caribbean is represented here, including blue holes and atolls. The region is synonymous with shark diving, and it is the splendid reefs with their marine life that support the many different species of shark.

Turks and Caicos to the south consists of around eight islands and a further 40 coral cays. These are situated on shallow banks atop a couple of ancient, massive volcanic peaks which are ringed by excellent coral reefs. Between the two banks are vertical drop-offs, between Providenciales (Provo) and West Caicos. The north-western side of Grand Turk has some spectacular walls cut with canyons and tunnels, which are best visited by dive boat.

MARINE CONSERVATION IN THE CARIBBEAN AND GULF ZONE

There are a number of different conservation initiatives throughout the Caribbean, all of which protect either specific habitats or particular groups of animals. One of these is the Specially Protected Areas and Wildlife (SPAW) Protocol instrumented by the Caribbean Biodiversity Fund (CBF). More than 100 million people benefit directly from the Caribbean's coral reefs, and more than 40 million are entirely dependent on these reefs for their livelihood and sustenance. Sadly, living corals have decreased around 50 per cent in the last 30 years, but as such are still a vital resource as part of an island nation's GDP.

The CBF was established in 2012 as a long-term funding source to protect the biodiversity of the Caribbean for future generations. It supports innovative solutions to advance conservation policies, and ensures that these are sustainable and manageable. Protecting the biodiversity of the Caribbean is a threefold plan, involving management and conservation of the seagrass beds, the mangroves and the coral reefs. This mixture of high-density habitats forms one of the world's most biologically diverse and productive ecosystems – so much so, in fact, that in 2016 the World Bank placed the economic value of the Caribbean's coastal and marine ecosystems at more than $54 billion.

Protection of this complex association is vital to allow this natural barrier to reduce the impacts of hurricanes and flooding.

To remove mangroves would double coastal flooding, triple the cost of damages, and remove the complex way that mangroves filter and purify coastal waters. Coral reefs sustain the Caribbean's beautiful white sandy beaches. Seagrasses in the region store about 1.3 billion tonnes of CO_2, according to a study by the University of Michigan. Current studies suggest that mangroves and coastal wetlands sequester carbon at a rate ten times higher than mature tropical forests. They also store three to five times more carbon per equivalent area than tropical forests.

Virtually all of the conservation zones are Marine Protected Areas (MPAs), and with around 600 sites covering over 50,000 km^2 (19,300 square miles), marine conservation is so closely linked with tourism that all countries in the Caribbean take marine conservation initiatives very seriously.

▶ Mangroves are vital for coastal communities, yet are also susceptible to extreme hurricane damage and pollution.

▲ Hawksbill Turtles are the most common species found in the Caribbean and are unafraid of any diver interaction.

NOAA states that MPAs 'vary widely in purpose, legal authorities, management agency, administrative approaches, level of protection and restrictions of human use'. Types of MPA include marine sanctuaries, national parks, national marine monuments, research reserves, natural reserves, wildlife refuges and seasonal closures. Despite this variation, MPAs generally have a common function: 'to conserve, manage and protect natural and cultural resources within marine environments.'

Although conservation efforts are usually only locally initiated to protect a vital area or resource, these small conservation areas are vital to the wellbeing of the entire region as marine creatures don't respect national boundaries. Plankton blooms and mass spawning occurrences spread their wealth of marine life not only throughout the Caribbean but also along the coast of the eastern United States and out as far as Bermuda and beyond. The coral reef connections are most obvious when the corals simultaneously release eggs and sperm into the ocean's currents, producing millions of tiny polyps ready to begin life, perhaps on the same reef system or perhaps many hundreds of miles away. Therefore small conservation successes provide a better understanding of how the entire region is interlinked, and how important it is for governments to cooperate in the wider scheme of international conservation.

The United States National Marine Sanctuary of the Flower Garden Banks in the Gulf of Mexico was established in 1992, when research demonstrated that these reefs have been building over the last 15,000 years. It took 13 years of work before President George H. W. Bush declared the area the US's 10th National Marine Sanctuary.

The Exuma Cays Land and Sea Park is a protected area covering a 455km² (176 square mile) stretch of the Exuma Cays. Established in 1958, it was the world's first land and sea park. The Bonaire National Marine Park extends to about 27km² (10.4 square miles). Established in 1979, it includes all the waters surrounding Bonaire and Klein Bonaire, from the high-tide mark to 60m (200ft) of depth.

In Cuba, the Jardines de la Reina archipelago of more than 250 virgin coral islands is located off the southern coast of Cuba. The marine reserve was established in 1996. It covers a large swathe of the archipelago and is one of Cuba's largest protected areas, with an area of 2,170 km^2 (840 square miles).

Dominica's Cabrits National Park is located on a peninsula on the northern end of the island, just beyond the town of Portsmouth. It was established in 1986, and while much of its area is terrestrial, the coastal areas around the reserve are also protected.

The Saba National Marine Park is one of the few self-sustaining marine parks in the world. A zoning plan divides the park for various recreational and commercial uses. It encompasses the waters and seabed encircling the island from the high-water mark to 60m (197ft) depth and was created in 1987.

Close to Saba is the Sint Eustatius National Marine Park, which was created in 1966. It encircles the Caribbean island of Sint Eustatius and extends up to 3km (1.8 miles) offshore. The Park protects habitats, including pristine, biologically diverse coral reefs, seagrass beds, sandy seabed, and open ocean communities, across an area of approximately 27.5km^2 (10.6 square miles).

Nearby, Sint Maarten's Man of War Shoal National Marine Park is located off the island's southern shore and is a sanctuary for whales, dolphins, sharks, turtles and many more marine animals. Established in 2010 as the country's first legally protected area, it covers 31km^2 (12 square miles) in an area also known as the Proselyte Reef Complex.

Established in 2001, the Virgin Islands Coral Reef National Monument is a US National Monument and consists of 5.1km^2 (19.8 square miles) of submerged lands within the 5km (3 mile) belt off Saint John, including Hurricane Hole and areas north and south of Saint John.

In the south-eastern Caribbean, the Tobago Cays Marine Park is owned and operated by the Saint Vincent and the Grenadines government and consists of a 5.7km^2 (2.2 square miles) sand-bottom lagoon, which encompasses the five cays, the inhabited island of Mayreau and the 4km (2.5 mile) Horseshoe Reef. The Marine Park was listed as a regionally significant ecosystem under the SPAW Protocol in December 2014.

Another one of the SPAW MPAs is the Glover's Reef Marine Reserve, which was established in 1993 and included in the SPAW list in 2012. Located 45km (28 miles) off the Belize coast, it lies at the heart of the Mesoamerican Barrier Reef and comprises about 800 coral patch reefs across 350km^2 (135 square miles). This is the largest coral reef system in the western hemisphere. Glover's Reef is a critical nursery and feeding ground for turtles, as well as sharks, rays and numerous other fish species. One of the few true atolls in the Caribbean, it became a UNESCO World Heritage Site in 1996. There are eight MPAs in Belize and all are an integral part of the Belize/Mexico Barrier Reef, the most significant barrier reef in the northern hemisphere.

At the top of the Mesoamerican Barrier Reef is the Isla Cozumel Biosphere Reserve. More than 1,100 marine species have been recorded within this reserve. Cozumel's total surface area, both terrestrial and marine, is 1,346km^2 (520 square miles). In addition, the location incorporates two Ramsar sites (wetlands of international importance): Arrecifes de Cozumel National Park, and Manglares y Humedales del Norte de Isla Cozumel.

▼ There is a dizzying array of marine life to be found on submerged walls, such as this Giant Anemone.

CLASSIFICATION AND NOMENCLATURE

The subsections in this guidebook begin with the group's family name, or in some cases a higher taxonomic rank such as order or class. Each species within that group is then introduced with its scientific ('Latin') name. Common names are also given wherever possible, including frequently used alternatives, but with so many different countries and languages around the borders of this sea, it is almost impossible to choose a single useful common name for every species. Scientific names are universal, and carry extra information too as they reflect the species' taxonomic position to the best of our knowledge (which means that they will be updated as necessary in the light of new genetic research).

Correctly naming a species is vitally important for your own log-book records, and is essential for scientists and marine biologists studying flora and fauna now and in the future. The modern binomial system of nomenclature (a two-word name denoting genus and species) was developed by the Swedish naturalist Linnaeus, and dates from publication of his *Systema Natura*, a compendium of known living species at the time, in 1758 and subsequent years.

A scientific name comprises the species' genus (first word) and specific name (second word). Both words are italicised, and the genus name has an initial capital – for example, *Epinephelus striatus*. The fish that bears this scientific name may also be known as Nassau Grouper, Stripe Grouper, Striped Reef Grouper or another variant – illustrating the importance of the single and universal scientific name. Once you get into the habit of using scientific names you will soon discover how useful it is to describe species in a common nomenclature that can be understood by both the scientific community and enthusiastic amateurs.

▸ The Nassau Grouper is a vital resource on all Caribbean reefs.

KEY TO SYMBOLS

This key describes the symbols that appear at the head of each species description. The symbols give a quick guide to the habit, diet and habitat of each species.

HABIT

- single
- pairs
- groups
- schools

DIET

- algae
- plankton
- invertebrates
- fish
- mammals

HABITAT

- caves
- seagrass
- sand
- reefs
- pelagic
- surface waters

EXTERNAL FEATURES OF FISH

This diagram illustrates the main structures of a fish referred to in the species descriptions.

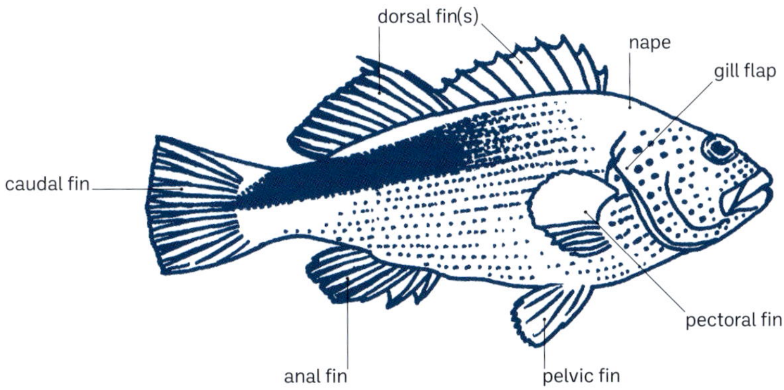

IDENTIFICATION GROUPS AND PICTORIAL GUIDE TO FAMILIES

Colour varies greatly between fish species, therefore it would seem an ideal means of identification. However, even within species, colour varies according to sex, age, region, season and surroundings. For this reason, body shape is a much more reliable means of identification. The following outlines represent most types of fish likely to be encountered in the Caribbean. Those sharing similar characteristics are grouped together for initial identification.

Sharks and rays

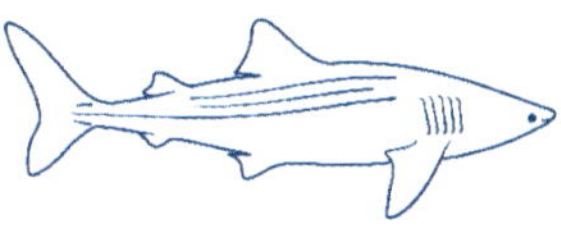

Whale Shark (*page 32*)

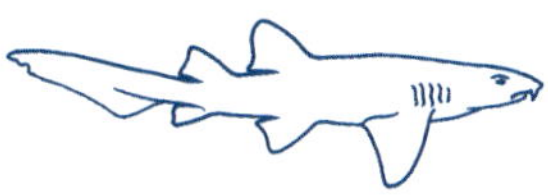

Nurse Shark (*page 32*)

Carcharhinid (requiem) sharks (*page 30*)

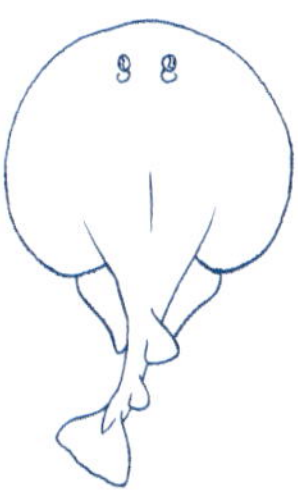

Electric Ray (*page 34*)

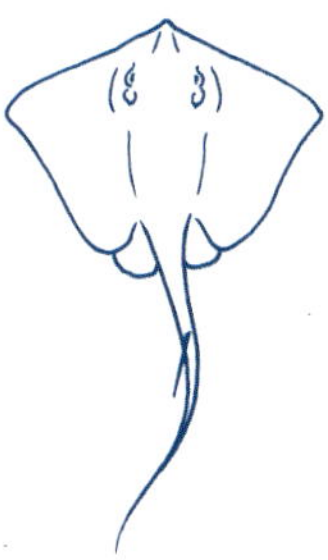

Stingrays (*page 34*)

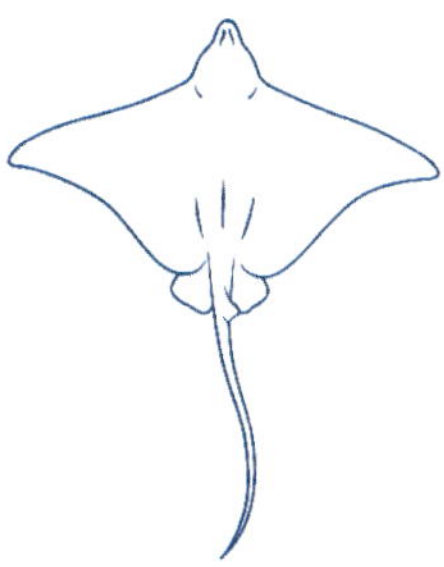

Eagle Ray (*page 34*)

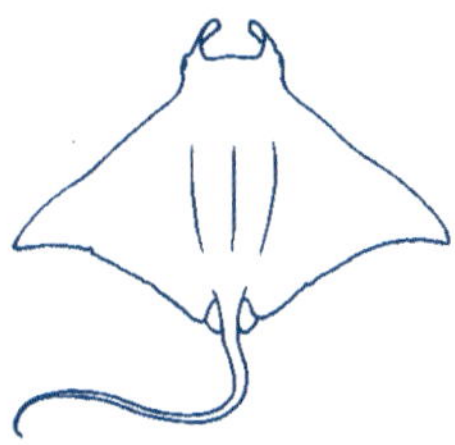

Manta Ray (*page 34*)

Silvery blade-shaped fish

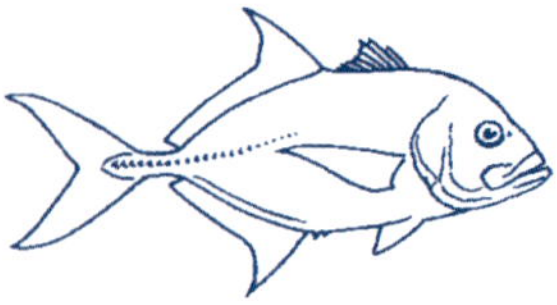

Jacks (*page 86*)

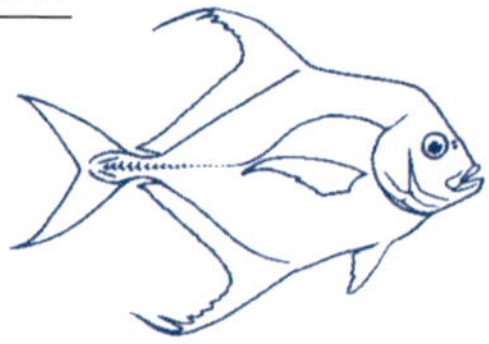

Palometa (*page 88*)

Silvery regular-shaped fish

Snappers (*page 102*)

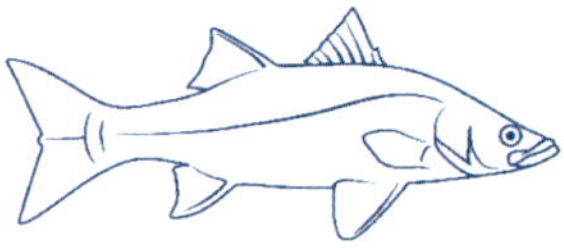

Snook (*page 90*)

Porgies (*page 94*)

Chub (*page 92*)

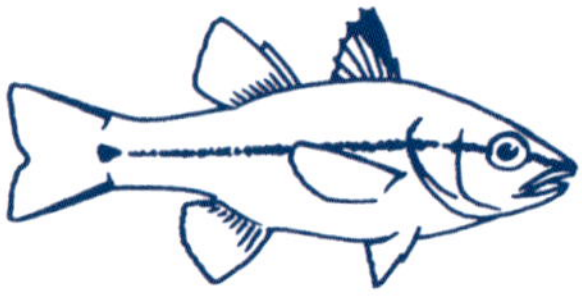

Cardinalfish (*page 74*)

Silvery torpedo-shaped fish

Tarpon (*page 90*)

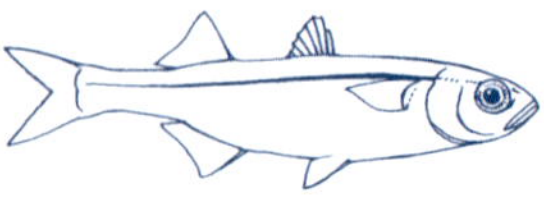

Silversides (*page 76*)

Silvery irregular-shaped fish

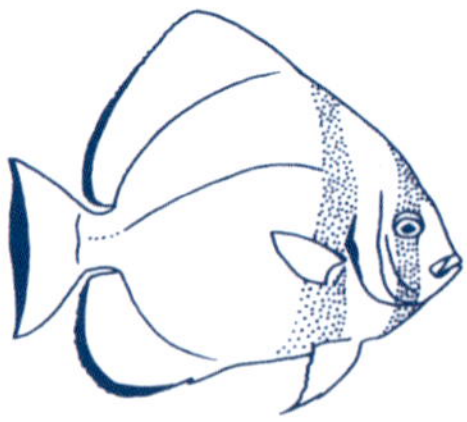

Spadefish (*page 92*)

Elongate fish above the reef

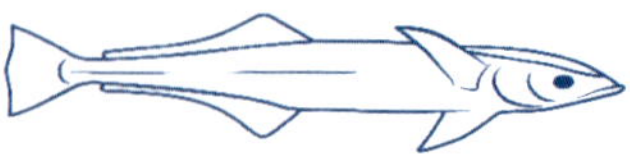

Remoras (*page 92*)

Barracuda (*page 90*)

Trumpetfish (*page 84*)

Elongate fish close to or on the seabed

Pipefish (*page 82*)

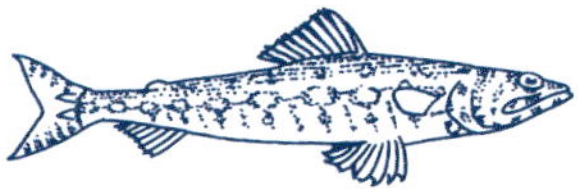

Lizardfish (*page 82*)

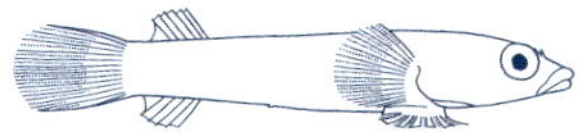

Clingfish (*page 70*)

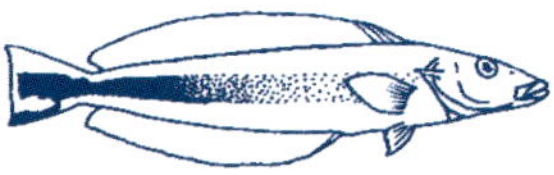

Tilefish (*page 84*)

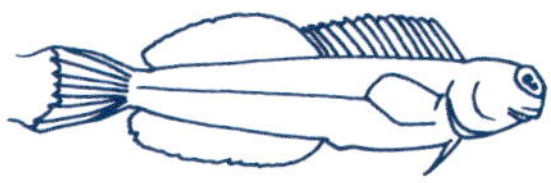

Blennies (*page 58*)

Gobies (*page 64*)

Moray eels (*page 36*)

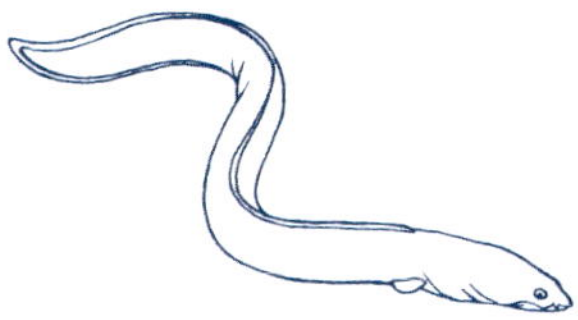

Sharptail Eel (*page 38*)

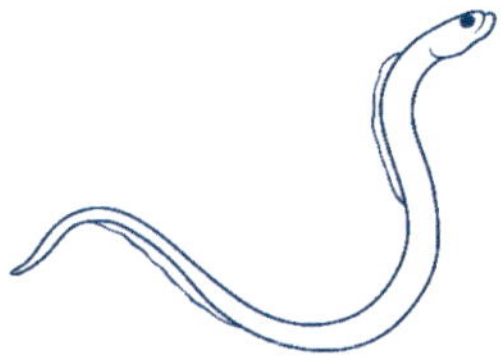

Garden Eel (*page 38*)

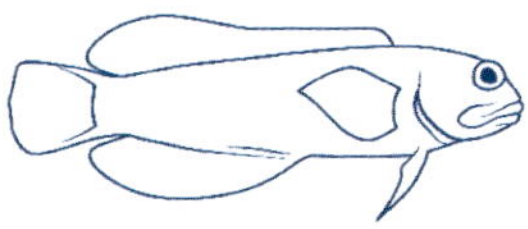

Jawfish (*page 70*)

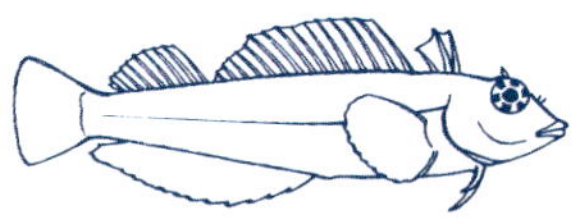

Triplefins (*page 64*)

Bigeyed/cave-dwelling fish

Soldierfish (*page 72*)

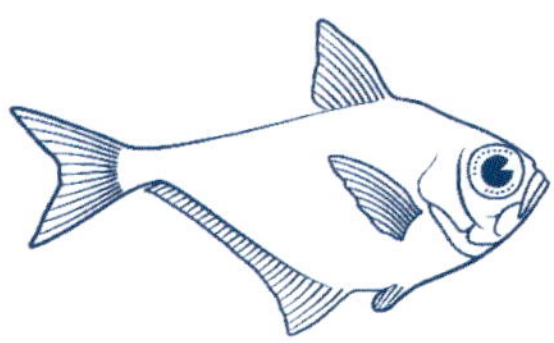

Sweeper (*page 76*)

Drum (*page 46*)

Squirrelfish (*page 72*)

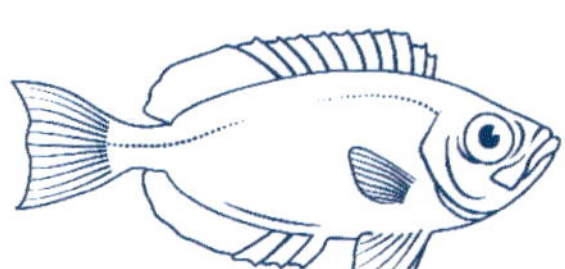

Bigeye (*page 72*)

Irregular-shaped fish close to or on the seabed

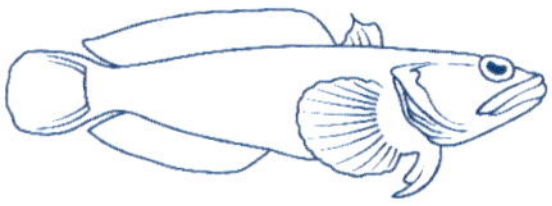

Toadfish (*page 80*)

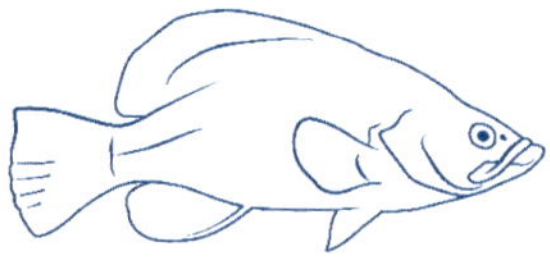

Soapfish (*page 56*)

Scorpionfish (*page 76*)

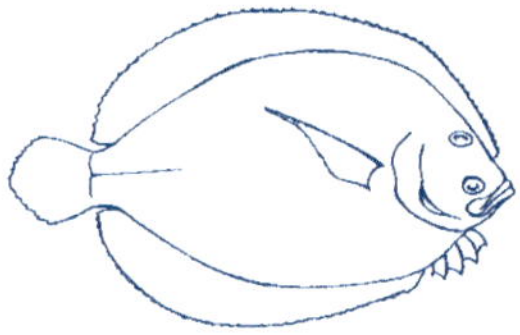

Flounders (*page 58*)

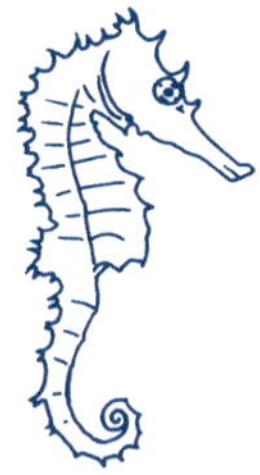

Seahorse (*page 82*)

Lionfish (*page 78*)

Frogfish (*page 78*)

Regular-shaped/swim with pectorals

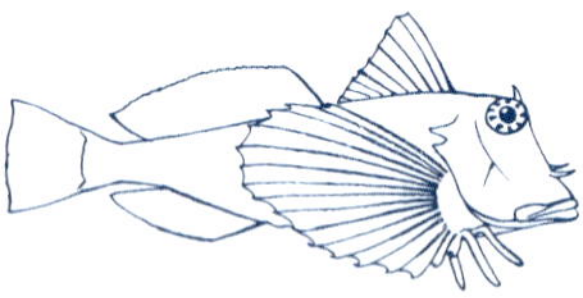

Gurnard (*page 80*)

Wrasse (*page 110*)

Hogfish (*page 110*)

Parrotfish (*page 106*)

Colourful oval-shaped fish

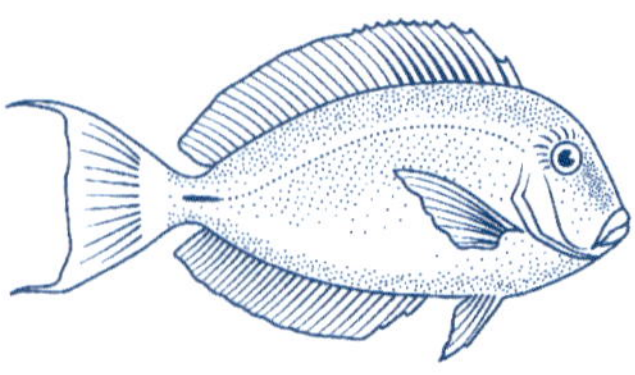

Surgeonfish (*page 100*)

Butterflyfish (*page 40*)

Colourful regular-shaped fish

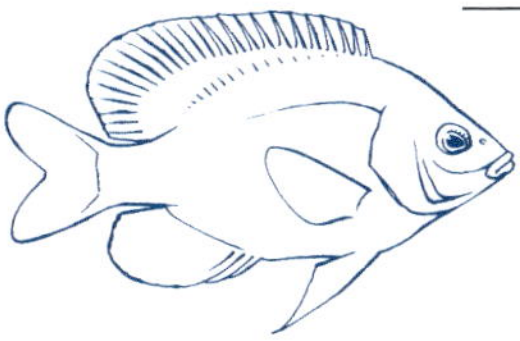

Damselfish (*page 42*)

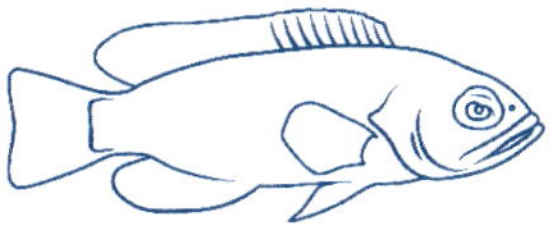

Basslets (*page 56*)

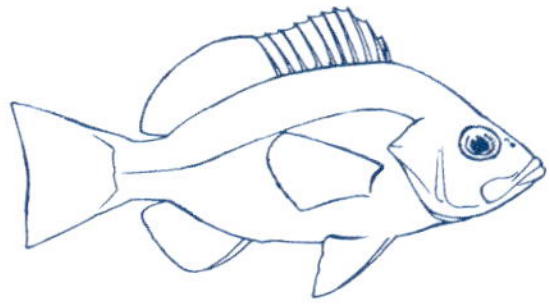

Hamlets (*page 52*)

Groupers (*page 48*)

Creolefish (*page 50*)

Hawkfish (*page 56*)

Goatfish (*page 94*)

Colourful irregular-shaped fish

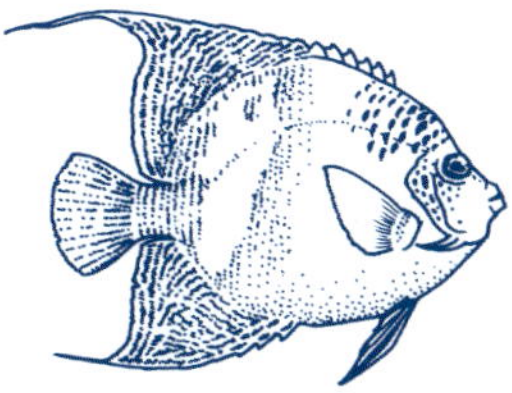

Angelfish (*page 38*)

Trunkfish (*page 98*)

Pufferfish (*page 96*)

Porcupinefish (*page 96*)

Triggerfish (*page 100*)

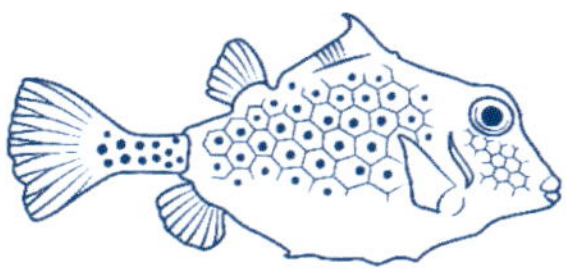

Cowfish (*page 98*)

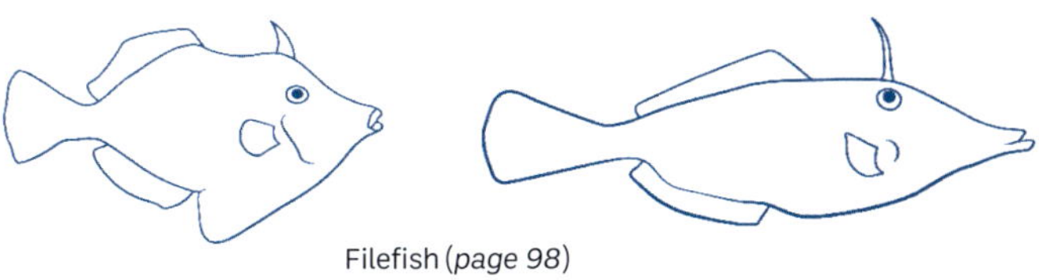

Filefish (*page 98*)

FISH

CARTILAGINOUS FISH (SHARKS AND RAYS)

THERE ARE TWO fundamental taxonomic groupings of fish: Chondrichthyes, which have cartilaginous skeletons and include all the sharks and rays, and Osteichthyes, the bony fish. Sharks and rays form a highly specialised and ancient group.

Sharks

Families: Rhincodontidae, Sphyrnidae, Carcharhinidae, Odontaspididae

ALL SHARKS EAT LIVE FOOD, but also scavenge decaying matter. Most are active nocturnal hunters, and all are shy. They use their highly developed sensory ampullae to detect low-level electrical currents and find sleeping fish.

1 GREAT HAMMERHEAD SHARK

Sphyrna mokarran

This pelagic shark is an opportunistic hunter, mainly feeding on seabed detritus but also hunting rays, pinning the ray down by its wide cephalofoil or 'hammer' and eating it from above. It is not known to be dangerous to humans, and amazing encounters may be experienced off Bimini in the Bahamas. Two other species of hammerhead shark occur in the region: the Bonnethead (*S. tiburo*) and the Scalloped Hammerhead (*S. leweni*).

2 TIGER SHARK

Galeocerdo cuvier

This shark is a fairly uniform grey with a white belly, and has dark stripes down its body which fade with age. It grows to around 5m (17ft). Tiger Sharks take all kinds of prey and are usually solitary hunters, although small groups will come together where food is abundant. To the north of Grand Bahama Island you can dive with them at a shallow reef now called 'Tiger Beach'.

3 BULL SHARK

Carcharhinus leucas

Quite a rotund shark, the Bull Shark can grow to around 4m (13ft) and is grey on top with a pale underside. It is considered quite an aggressive species, and because it lives in warm shallow waters and even far up freshwater rivers, attacks on humans do occur occasionally. This, of course, means that it is killed indiscriminately and consequently has declined and is considered Vulnerable by the IUCN.

4 SILKY SHARK

Carcharhinus falciformis

This is a pelagic shark, growing to around 3m (10ft) and usually encountered off the offshore reefs and small islands. Often found in small hunting packs, it is quite curious and may approach closely.

1

2

3

4

1 SANDBAR SHARK

Carcharhinus plumbeus

This species grows to around 1.8m (5ft 11in). It has a brownish-grey body with a lighter underside, and a very obvious large triangular dorsal fin. It prefers shallower coastal zones over sand flats. This is a common shark with a commercial market for its palatable meat, and is known to live quite happily in large aquarium settings.

2 BLACKTIP SHARK

Carcharhinus limbatus

This uncommon shark grows to 1.5m (4ft 11in). It is very streamlined but stout, with a pointed dorsal fin, a long pointed snout, and obvious black tips to all of its fins. It prefers a more pelagic existence, but will patrol the reefs in search of fish and carrion.

3 CARIBBEAN REEF SHARK

Carcharhinus perezi

The most common of all the shark species found in the Caribbean, this species grows to around 3m (10ft). Once highly valued for its meat and liver, it is now regarded as being much more valuable for the ecotourism trade. It is unafraid of people, and divers are able to hand-feed this species in a number of areas around the Caribbean. Care should be taken as always, as other shark species not usually associated with the shark-feeding circus may stray into the area and can be quite aggressive. Whilst some Caribbean islands embrace this practice, others are rightly concerned over the impact that these encounters have in habituating sharks to associate people with food.

4 LEMON SHARK

Negaprion brevirostris

Named for its yellow coloration, to aid its camouflage when swimming over sand flats, this species grows to around 3m (10ft). It has a flattened head and a short, wide snout. Years of research have been undertaken on this species by the Bimini Shark Lab, where these sharks are prevalent and have their young amidst the mangrove and seagrass beds.

5 THRESHER SHARK

Alopias vulpinus

Quite rare in the Caribbean, this species is more common around some of the southern islands of the Bahamas. Preferring deeper water and offshore reefs, it is a dark blue-grey above, with a white underside, and looks almost metallic. It has a very rounded head, large dorsal fins and an obvious elongated upper lobe to its tail, and can grow to around 5–6m (17–20ft).

6 WHALE SHARK

Rhincodon typus

The largest fish in the sea, growing to around 12m (39ft), the Whale Shark is a plankton feeder. Although generally rare in these waters, it is regularly seen off the north coast of the Yucatán and in the southern region around the island of Utila in the Dutch Antilles. It is always accompanied by many different fish, especially remoras or suckerfish. Each individual has a unique pattern of spots and blotches.

7 NURSE SHARK

Ginglymostoma cirratum

This common, brownish-coloured shark is easily identified as all of its fins are rounded and the caudal fin (tail) is elongated. It has a broad head with a pair of barbels above the jaw. A very sedentary shark, it is most often encountered hiding under overhangs or in the 'grooves' of a spur and groove reef, and is found all over the Caribbean and Gulf of Mexico, as well as South America.

1

2

3

4

5

6

7

Rays

Families: Torpedinidae, Rajidae, Dasyatidae, Myliobatidae, Mobulidae, Urolophidae

RAYS ARE INSTANTLY recognisable as their pectoral fins are joined onto the head and form a broad skirt around the body. They have no dorsal fins and their mouthparts are on the underside of the head.

1 LESSER ELECTRIC RAY

Narcine brasiliensis

This is a small ray, growing to around 45cm (1ft 6in) long. It has an oval, elongated body with a round snout and a triangular caudal fin. With its jaws moving as a unit, it sucks in its prey of small fish and crustaceans. Its electrical organs can generate an average voltage of about 0.35mV, which is useful for stunning prey, but is more effective as a form of defence against attack from above.

2 YELLOW STINGRAY

Urolophus jamaicensis

This small and colourful ray grows to around 60cm (2ft). With its yellow and creamy-brown blotched back, it is quite unmistakable. Its skirt is rounded, as are the pelvic fins, and it has a venomous spine near the tail. Usually found on the seabed and amidst the coral rubble near larger reefs, it prefers shallower waters of less than 18m (60ft) and is reasonably common throughout the Caribbean.

3 SOUTHERN STINGRAY

Hypanus americanus

This is by far the most common stingray throughout the entire region. It has huge pectoral fins which can grow to a diameter of more than 2m (6ft 6in); it uses the fins to stir up the sand in search of prey, or to bury itself when settling onto a sandy bottom. The Southern Stingrays at the famous 'Stingray City' on northern Grand Cayman Island are well used to human interaction and in most cases can be handled, with care.

4 ROUGHTAIL STINGRAY

Bathytoshia centroura

This species is larger than the Southern Stingray at over 2.6m (8ft 6in) across, and 360kg (800lb) in weight. Another bottom-dweller, it is dark grey with a pale underside. The long tail is about 2.5 times the length of the body disc and its name comes from the saw-toothed spinal ridge that sits atop its tail.

5 GIANT MANTA RAY

Mobula birostris

The largest of the rays in the Caribbean and Gulf of Mexico, the Giant Manta Ray has a prominent pair of cephalic fins just above the jaw, which it extends and furls over to help scoop plankton into its cavernous mouth. Growing to over 7m (23ft) in width, it is blue-black on top with a white underside that usually has identifying dark blotches. Often accompanied by remoras, it swims very gracefully and is a huge delight to see.

6 SPOTTED EAGLE RAY

Aetobatus narinari

The Spotted Eagle Ray has an obvious snout with which it digs up the seabed in search for molluscs and crustaceans. It often comes fairly close to shore, and will be encountered on many dives, particularly on the sand flats or inner lagoons. It is common in all areas of the Caribbean and north-western Atlantic, from Bermuda to virtually every part of the globe within the tropics.

1

2

3

4

5

6

BONY FISH

MORE THAN 90 PER CENT of all other fish in our seas, rivers and lakes are in this group. They have bony rather than cartilaginous skeletons, and they all have a large bony plate that protects the gills. Most species fall into the taxonomic class Actinoptergii, or ray-finned fish, and are characterised by distinct rays or spines on their fins. The remainder form the class Sarcopterygii, the lobe-finned fish. There are approximately 1,000 species of bony fish in the Caribbean, of which more than half live around the coral reefs.

Eels

Families: Muraenidae, Ophichthidae, Congridae

THERE ARE MANY SPECIES of moray eels in the Caribbean, but not all occur in the same areas at the same time. They are long and sinuous and have continuous dorsal and anal fins which run the entire length of the body. Most are associated with the coral reef, but stay hidden in holes and come out to scavenge for food at night, principally preying on sleeping reef fish. The upper jaw has three rows of sharp pointed teeth, the third row in the upper centre of the jaw. They always look threatening due to the opening and closing of the jaws; however, this reflex is only to aid respiration.

1 GREEN MORAY

Gymnothorax funebris

The largest of the moray eels, this species grows to over 1.8m (5ft 11in) in length. It is olive-green to greenish brown, with no visible markings. and is an active predator on the reef in late afternoon and at night. Although it appears unconcerned by divers, this is only due to its poor eyesight! It is found throughout the entire region.

2 SPOTTED MORAY

Gymnothorax moringa

Common throughout the eastern and western Atlantic Ocean and throughout the Caribbean zone, this distinctive eel grows to around 120cm (4ft). It is pale grey, and covered completely in brownish spots and blotches. Usually found in depths less than 12m (39ft), it forages at all times of the day and night.

3 GOLDENTAIL MORAY

Gymnothorax miliaris

A small eel, only growing to around 60cm (2ft), this species is brownish and covered with small yellow spots. As the name suggests, it has a yellow tip to its tail, and also has a golden yellow ring around the black pupil. This species also forages during the day and is found throughout the region, principally on coral rubble and algae-rich areas.

4 CHAIN MORAY

Echidna catenate

This fairly stout eel rarely exceeds 60cm (2ft) in length. It is distinctly dark, with whitish-cream chain-like markings, often linked together. Usually the head protrudes from its hole in the reef, and it actively forages at night for small crustaceans and sleeping fish. It is found in shallow waters from Bermuda to Brazil.

5 VIPER MORAY

Enchelychore nigricans

This brown-coloured moray eel has an upturned upper lip, midway along the jaw, exposing a mouth full of needle-sharp teeth. It often shares its home in the reef with numerous cleaner shrimps. It is quite shy, although you can usually see the head with its long nostrils quite clearly before it retreats. It forages at night and is found on the deeper reef.

1

2

3

4

5

1 SHARPTAIL EEL

Myrichthys breviceps

Although it looks like a moray eel, this species is in a different family. It grows to around 120cm (3ft 11in) and has downward-pointing nostrils just above its fleshy lips. It is greyish-olive with small cream spots on the head and large cream spots all over the rest of the body. It uses a wide range of habitats; I have found it near eelgrass beds at the edge of shallow coral reefs.

2 BROWN GARDEN EEL

Heteroconger longissimus

The most common garden eel in the Caribbean and a member of the conger eel family, this species can grow to around 1m (3ft 3in). It lives in deep vertical burrows in fine sand, and feeds by extending its body out into the current to catch plankton. It will reverse tail first into its burrow should you get too close. It is usually found in large groups, swaying in the current.

3 BLACKSPOTTED EEL

Quassiremus productus

This large eel grows to over 1m (3ft 3in). It spends the day well hidden under the sand with only its snout visible, only venturing out at night. Unfortunately, this makes identification very difficult, as the body markings help determine the species. It is considered rare in this region; my specimen was found on a night dive off Dominica.

Angelfish

Family: Pomacanthidae

ANGELFISH ARE AMONG the most colourful fish in the Caribbean and occur throughout the entire range. Feeding on algae and sponges, they perform a vital function on the coral reef in limiting the growth of smothering algae. Some also feed on sea fans, and juveniles have been seen to act as 'cleaners' on other fish species.

4 QUEEN ANGELFISH

Holacanthus ciliaris

This is the most colourful of all the angelfish to be found in the Caribbean; even the juveniles are incredibly garish. It is instantly recognisable by the blue and yellow coloration and the obvious 'crown' on the nape. It feeds on sponges and, when breeding, one male may have several females in his harem. Found throughout the western Atlantic and Caribbean, it spends its life amidst the coral reefs and never in open water.

Juv. 4

5 BLUE ANGELFISH

Holacanthus bermudensis

Closely related to the Queen Angelfish is the Blue or Bermuda Angelfish. It is a paler, more uniform blue, and lacks the obvious crown found on the forehead or nape of the Queen Angelfish. The two species are known to interbreed and a new separate hybrid species is now recognised.

6 TOWNSEND ANGELFISH (HYBRID)

Holacanthus townsendii

A hybrid species between the Queen and Blue Angelfish, this angelfish has an indistinct crown, and is mainly golden-coloured with a yellow tail and blue highlights around the edge of the body, eyebrow and gill slits.

1
2
3
4
5
6

1 ROCK BEAUTY

Holacanthus tricolor

Unmistakable in colour, the Rock Beauty is black with a brilliant yellow head and fins, and blue eyes. When young it looks very similar to a damselfish, but has a large black spot in the centre of the body, that gradually expands to cover the entire body as it matures. It is extremely shy and always attempts to keep out of sight. It occurs throughout the Caribbean.

2 FRENCH ANGELFISH

Pomacanthus paru

A large angelfish, growing to around 40cm (1ft 4in). It is laterally compressed and almost circular in shape, with a steeply sloping head, small snout and a pointed white mouth. Coloured an almost metallic blue-black, it has yellow highlights to its scales and around the eye. Juveniles are black with golden yellow stripes.

3 GREY ANGELFISH

Pomacanthus arcuatus

This angelfish is grey-brown with a small blue line on the edge of its tail. Fairly common throughout the region, adults are found in pairs and are often associated with French Angelfish.

Butterflyfish

Family: Chaetodonidae

BUTTERFLYFISH ARE SHAPED somewhat like angelfish, with a compressed deep body, but have a much more pointed snout with a small mouth, perfect for probing amongst the corals and coral debris in search of small invertebrates and coral polyps. They also often have a facial stripe that disguises their eyes, or even have markings at the tail end that mimic an additional eye! They are usually found in their life-long mating pairs.

4 BANDED BUTTERFLYFISH

Chaetodon striatus

One of the more common species, the Banded Butterflyfish is around 15cm (6in) long. It has a dark stripe running vertically through the eye, two additional vertical bands on the body, and black edges to the fins and tail. Always in pairs, it often mixes with other butterflyfish and angelfish, and is found throughout the region, but always in shallower water of less than 18m (60ft). At night, the vibrant daytime colours fade away almost completely, except for the dark stripe through the eye and a dark spot just at the base of the tail, which helps to confuse predators.

5 FOUREYE BUTTERFLYFISH

Chaetodon capistratus

This distinctively coloured fish grows to around the same size as the Banded Butterflyfish. It is pale yellow and cream with pale grey chevron markings down the length of the body, a pale stripe which runs through the eye, and a very obvious large ocellus or false eye marking near the tail which is black and ringed in white, similar to its real eye (hence the name).

1

2

3

4

5

1 SPOTFIN BUTTERFLYFISH

Chaetodon ocellatus

Slightly larger than other butterflyfish at 20cm (8in) long, this species has a spot at the end of its dorsal fin. The body is silvery white with yellow on all fins and the tail. There is also a vertical black stripe through the eye. At night, its colours change dramatically, showing an additional black splotch near the tail and a wide grey diagonal stripe which runs the length of the body. It feeds on invertebrates and coral polyps.

2 REEF BUTTERFLYFISH

Chaetodon sedentarius

This species grows to around 18cm (7in). It is pale creamy yellow on top, fading to white below, and has a wide dark band at the rear which fades into the tail. It has yellow fins and tail, and a black bar through the eye. It is not often seen as it prefers deeper parts of the reef, around 15–40m (48–130ft). It is uncommon but widespread, occurring from the Carolinas to Mexico.

3 LONGSNOUT BUTTERFLYFISH

Chaetodon aculeatus

This is the smallest butterflyfish at around 9cm (3½in). Almost triangular in shape, it has a long pointed snout and a very small mouth. The body colour shades from almost black at the rear of the dorsal fin to a pale cream underneath the body and pelvic fins. It has no dark eye band, but shows a flash of mustard yellow above the eye. As with the rest of this colourful family, the front of the dorsal fin is separated into spines which it extends when threatened.

Damselfish and chromis

Family: Pomacentridae

DAMSELFISH ARE VERY COMMON throughout the Caribbean and various species will be seen on every dive. Quite often, both adults and the more colourful juveniles will be found at the same time. Many are omnivores, and vigorously guard a little 'garden' patch around an area of dead coral. Others, like the Sergeant Major, are always found in the water column and usually in quite shallow water, where they aggregate in large numbers and eat plankton.

4 THREESPOT DAMSELFISH

Stegastes planifrons

There are six species in the genus *Stegastes* found in the Caribbean. All enjoy relatively shallow waters of less than 18m (60ft), and all are around the same length of 10cm (4in). The Threespot, Beaugregory and Yellowtail Damselfish are very similar in their adult form. Juveniles of the Threespot are brilliant yellow with a black spot on the rear of the dorsal fin and another on the saddle of the upper base of the tail.

Juv. 4

5 BEAUGREGORY

Stegastes leucostictus

Beaugregory juveniles have yellow undersides and are electric blue on top, fading to blue spots, from the nape to the rear of the dorsal fin. They have no obvious dark spots. The adult retains its yellow tail and is usually lighter around the front of the head than the Threespot. The body is shaded darker, then blends into the yellow or tan tail.

1

2

3

4

5

1 BICOLOR DAMSELFISH

Stegastes partitus

The most distinctive *Stegastes* species, this fish is boldly two-tone with a black forebody and white rear, and a translucent white tail. Occasionally the forebody is itself two-tone, with a black upperside and yellow underside. It inhabits patch reefs and areas of coral debris in spur and groove reefs, where the male will guard the eggs vigorously.

2 COCOA DAMSELFISH

Stegastes variabilis

Like the Threespot Damselfish, the adult of this species is generally dark brown with a pale area marked with a black spot on the saddle of the tail. The underside is pale metallic tan. The juvenile steals the show, being brilliant blue from the crown to the rear of the dorsal fin, and brilliant yellow underneath.

3 DUSKY DAMSELFISH

Stegastes adustus

In this species, the dorsal and anal fins are rounded and extend over the saddle of the tail. The tail is rounded and forked, and all fins are edged in blue except the pectoral fin, which is ochre. The juvenile is greyish tan, with orange along the top of the dorsal fin. There is a black spot ringed in electric blue at the rear of the dorsal fin, and a row of brilliant blue spots leading from the eye to the tail. This fish can be found in all zones.

Juv. 3

4 LONGFIN DAMSELFISH

Stegastes diencaeus

The juvenile of this species is flamboyant in the extreme, being golden yellow below and vibrant blue above, in dotted stripes from the eye to the rear of the dorsal fin, which has a dark spot to the rear, like another eye. The adult is rather dull greyish brown to almost black.

5 YELLOWTAIL DAMSELFISH

Microspathodon chrysurus

This is another species in which the juvenile is especially colourful, being dark blue with lighter brilliant blue random spots everywhere except the tail, which is almost always a brilliant yellow (but can also be a translucent white). The adult is dark brownish blue and retains its blue spots, albeit now much smaller. It also retains its brilliant golden yellow tail.

6 NIGHT SERGEANT

Abudefduf taurus

Growing to around 25cm (10in) and very similar in shape to the more common Sergeant Major, the Night Sergeant prefers shallow open waters less than 6m (20ft). It occurs from southern Florida through the Bahamas and into the Caribbean. It is dull greenish grey with five darker vertical bands down the length of the body. The dorsal fin is slightly spiky on top and it has a rather pointed nose and mouth.

1

Juv. 2

3

Juv. 4

Juv. 5

6

1 SERGEANT MAJOR

Abudefduf saxatilis

This species can grow to around 17cm (7in). It has obvious dark silvery bands, with yellow tinges in between. Males become darker-coloured when guarding eggs, and are very aggressive if anything gets too close. It is often seen in large schools near the surface, feeding on plankton.

2 BLUE CHROMIS

Chromis cyanea

This species reaches a maximum of 10cm (4in) but is usually much smaller than this. It is brilliant blue overall with a darker blue upperside running from the nape to above the deeply forked tail. Found throughout the entire region and out to Bermuda, the Blue Chromis is usually in medium-sized, loose aggregations, feeding on plankton above or out from the reef edge. It prefers the lower reef, but rarely occurs below 25m (80ft). It will retreat into the reef if danger threatens.

3 SUNSHINEFISH

Chromis insolata

The Sunshinefish grows to a maximum of 10cm (4in). It is well named with its purple lower body, fins and tail and a sunshine-yellow upper body and dorsal fin. It prefers the deeper reef, from 15–60m (50–200ft), and is relatively common across the entire Caribbean and Gulf of Mexico.

4 BROWN CHROMIS

Chromis multilineata

This species grows to around 20cm (8in). It is olive-green with a yellow fringe to the dorsal fin and tip of the upper tail. It has a dark spot at the base of the pectoral fin. It is regarded as common throughout the entire region but not Bermuda. It can be found in loose aggregations around healthy corals, and ventures well out into the current to feed on plankton.

Drums

Family: Sciaenidae

DRUMS ARE FAIRLY COMMON on all reefs, usually being found under an overhang. The active 'dance' of the juveniles always give them away, due to their vivid black and white stripes and long fins.

5 SPOTTED DRUM

Equetus punctatus

The Spotted Drum can grow to a maximum of 30cm (1ft), only growing into its spots when it is adult. The spots are located on the dorsal fin and tail. The rest of the body bears black and white stripes that extend up the front of the first dorsal fin. This fin, and the tail, are especially long in juveniles, and there is a broad black band between the two. It is found on most reefs.

6 HIGHHAT

Pareques acuminatus

Shorter-bodied than the Spotted Drum at up to 20cm (8in), but stouter, this species is dark greyish brown with pale stripes along the length of the body, but not on the second dorsal fin or tail. Juveniles are similar to those of the Spotted Drum, but the stripes on the Spotted Drum are vertical, while those on the Highhat are horizontal.

1

2

3

4

5

6

Groupers

Family: Serranidae

THE GROUPER FAMILY has several representatives in the region. Many have been hunted mercilessly for many years by fishermen. The flesh is highly prized, and when you consider that larger specimens can weigh several hundred kilos, it is clear why they are so valued. Thankfully, there are many established marine parks where groupers are still abundant and are commonly seen on every dive. Members of this family begin life as female and change to male as they get older. They are known to mass spawn.

1 NASSAU GROUPER

Epinephelus striatus

This is one of the most common grouper species on shallow reefs. It grows to around 1m (3ft 2in). It has large brown longitudinal stripes above the eyes, blending into vertical stripes at the rear of the body. The nostrils are large and pink, and it has fleshy lips that it extends to catching prey and to allow cleaning wrasse and gobies to enter the mouth. Its very inquisitive nature means that it will accompany divers as they explore the reefs and will dart in and catch fish that are disturbed; they particularly like squirrelfish!

2 GOLIATH GROUPER

Epinephelus itajara

The largest grouper, this species grows to around 2.5m (8ft 2in). It has large fleshy lips and the dorsal fin rays are separate at the tips. It has brown and grey mottled markings against a fawn background, and is well camouflaged when it ventures into shallower water in the search for food. However, it generally inhabits deeper reefs, caverns and large shipwrecks, where there are plenty of places for it to hide. Larger specimens will be accompanied by remoras.

3 RED GROUPER

Epinephelus morio

Growing to around 1.25m (4ft), this fairly large grouper is reddish brown, often with white spots towards the rear of the body. It has two separate nostril holes below each eye. It is uncommon on the lower reef and often rests on the bottom. It has a patchy distribution but occurs in Bermuda, the Carolinas, throughout the Caribbean and Gulf of Mexico and south to Brazil.

4 RED HIND

Epinephelus guttatus

This smaller grouper grows to around 76cm (2ft 6in). It is light-coloured, but completely covered in red and brownish red spots, sometimes merging into blotches. Its fins have black margins, except the dorsal fin, where the forward fin rays may be tipped yellow. It is common on most reefs.

5 MARBLED GROUPER

Dermatolepis inermis

The Marbled Grouper grows to around 76cm (2ft 6in), and has very distinctive markings. Juveniles are brown with intermittent white blotches, while adults are white with brown and grey marbling all over the entire body and fins. It has a high back and steep nape, leading to a relatively small mouth. It is usually found on the lower reef, below 21m (70ft).

6 GRAYSBY

Cephalopholis cruentata

The Graysby, although similar to the Red Hind, is paler overall and has three to five pale or dark spots along the base of the dorsal fin. It reaches 30cm (1ft) in length and is common on most Caribbean reefs. It prefers shallower waters over good, healthy reefs above 18m (60ft) and can be approached fairly easily if you go slowly. Juveniles have a very obvious white stripe over the lips and leading down the nape.

1

2

3

4

5

6

1 CONEY

Cephalopholis fulva

Identification of the Coney is always difficult, due to its many colour changes and different colour variants. It can be almost identical to the Red Hind except that the entire body is red with a covering in dark spots, and darker fins, tail and underside. A bicoloured variant has a red upperside and white underside, scattered with tiny black or blue spots. A golden variant has a number of blue spots around the snout and under the eye, and greenish eyebrows. Juveniles tend to have small black and white stripes towards the rear of the dorsal fin.

2 BLACK GROUPER

Mycteroperca bonaci

A large grouper, growing to around 1.2m (4ft), this species also occurs in a few colour variations, most typically pale and mottled with a black fringe to its fins and tail, or entirely black. It prefers the mid reef and reef crest, before the wall plummets into the depths, but can also be found around the 30m (100ft) range.

3 TIGER GROUPER

Mycteroperca tigris

This grouper is quite long and slim, growing to around 1m (3ft 3in). It has around nine prominent wide brown or pale stripes over a lighter body. It tends to look fiercer than some of the other grouper species and has the ability to change colour quite dramatically, depending on the environment and its stress levels. It can often be found in cleaning stations where it may just rest amongst the corals. Juveniles are greenish yellow with a dark brown horizontal stripe running the length of the body that fades with age.

4 YELLOWFIN GROUPER

Mycteroperca venenosa

This grouper grows to around 90cm (3ft). It can vary in colour; it usually has grey to red mottling over a paler background, but can also change colour rapidly. It has obvious yellow margins to its pectoral fins. It occurs around the shallower reef crest, where it is often seen hiding amongst soft corals.

5 CREOLEFISH

Paranthias furcifer

This fish is always tan-coloured, with three spots down its back and a forked tail. It can grow to 30cm (1ft) and is an active plankton feeder on steeply sloping reefs throughout the region. The juvenile looks similar to a juvenile Graysby, with a white stripe that reaches from its lips up the nape to the dorsal fin.

Bass

Family: Serranidae

6 HARLEQUIN BASS

Serranus tigrinus

This bass has distinctive dark, tiger-like stripes down the length of the body and a silvery yellow underside. The markings also include two rows of dark spots running the length of the body, and a dark spot at the front of the dorsal fin. The head is much more pointed than in other species in this family. Often found in pairs, it inhabits shallow waters with coral rubble, where it hunts for small crustaceans and other invertebrates. It grows to around 10cm (4in) long.

1

2

Juv. 3

3

4

5

6

1 LANTERN SEA BASS

Serranus baldwini

This species only grows to around 5cm (2in). It has dark reddish-brown splotches and spots, with more red around the snout and a red-ringed eye. Its fins are quite translucent and its lower flanks bear a series of four almost rectangular dark splotches with red underneath. A shy fish, it is common in coastal waters.

2 BELTED SANDFISH

Serranus subligarius

Similar in shape to other bass, and growing to 10cm (4in), the Belted Sandfish is more common in the northern Caribbean, but does extend from the Carolinas south to Brazil. A beautiful small fish, it is patterned with brown spots, has a white belly and intricate black markings on the rear of the body, extending to the rear of the dorsal fin and tail.

3 TWINSPOT BASS

Serranus flaviventris

Growing to just over 5cm (2in), this species is pale creamy white or pale blue with two lateral lines running the length of the body, one from the upper lip to the mid-body and the other from above from the nape to midway along the underside of the dorsal fin; this second line may have a series of spots along the length.

4 TOBACCOFISH

Serranus tabacarius

A larger bass, growing to around 15cm (6in), the Tobaccofish is identified by a broad ochre stripe that runs the entire length of the body. It has dark stripes around the head, one of which runs through the eyes and around the snout. Tending to stay around the bottom of coral rubble areas in search of invertebrates, it is relatively common throughout the Caribbean.

5 SAND PERCH

Diplectrum formosum

The Sand Perch is another of the larger bass species, growing to a maximum of 30cm (1ft). It is a long, streamlined fish with a series of obvious, sometimes blue-tinged stripes along the upper half of the body. Whilst it is rarer in the greater expanse of the Caribbean, it occurs as far west as Cuba and the Cayman Islands, but prefers the eastern Caribbean, from Bermuda to Uruguay.

Hamlets

Family: Serranidae

SMALL MEMBERS OF the grouper family, hamlets are colourful and curious, yet shy. Most species are widely spread throughout the entire region, but only one species is recorded from Bermuda, the Barred Hamlet. Known only from the western Atlantic and Caribbean, this hamlet's taxonomy is controversial, with some biologists defining it as one species with around ten colour variations, but others splitting some or all of the variants into distinct species.

6 BUTTER HAMLET

Hypoplectrus unicolor

This species can grow to around 15cm (6in). It is silvery white, blending to pale yellow fins, and has an obvious dark spot just at the juncture of the tail. The tail itself is almost translucent and rounded. The snout is pale yellow and there may be pale blue stripes radiating around the mouth. Hybrids may have additional black spots on the tail and snout; these markings may be outlined in blue. It prefers shallower water, around 18m (60ft).

1
2
3
4
5
6

1 BARRED HAMLET

Hypoplectrus puella

The most common of the hamlets, this is arguably the species from which the largest number of hybrids are derived. It grows to around 17cm (7in). Its colours are variable, but most have pronounced wide, vertical, brown stripes down the body and may have a black spot at the tail. The central stripe takes up most of the body; this fades to cream on the underside. It occurs throughout the entire region.

2 TAN HAMLET

Hypoplectrus randallorum

This fish is uniform tan with a series of small spots at the start of the anal fin or tail. These spots fade with age. Now that we understand that hamlets breed in mating pairs or threes and may interbreed, there will probably be more hybrid forms discovered.

3 GOLDEN HAMLET

Hypoplectrus gummigutta

This hamlet is golden yellow, including all of the fins and tail. It has a very distinctive black irregular blotch either side of the mouth, outlined in bright blue. This same neon blue colour is repeated as spots and lines around the eye and continuing around the mouth and behind the eye. Not usually found below 30m (100ft), it is more common in the central Caribbean around the Cayman Islands and southern Cuba.

4 SHY HAMLET

Hypoplectrus guttavarius

Very similar to the Golden Hamlet, this fish has a mainly black body, fading to golden yellow at the nape and snout, as well as the tail and all other fins. It also has black splotches either side of the snout, ringed in neon blue, and thinly outlined in black.

5 INDIGO HAMLET

Hypoplectrus indigo

This species grows to around 17cm (7in). It is very similar to the Barred Hamlet, but the lighter patches are pale grey-blue and the darker vertical stripes are vivid blue, with the same shade on the fins, tail and snout. A darker stripe runs through the eye; here there is also a white slash with a darker stripe set within. It is locally common in a number of areas.

6 BLACK HAMLET

Hypoplectrus nigricans

Very distinctive, but sometimes difficult to spot, this species grows to around 17cm (7in). It is entirely dark charcoal grey with no additional markings. Usually found on the lower reef or reef crest in waters deeper than 18m (60ft), this is a very shy fish, but is curious enough to allow you to get close enough to photograph.

7 MASKED HAMLET

Hypoplectrus providencianus

Growing to just over 10cm (4in), this hamlet is considered uncommon throughout the central Caribbean, but occasionally found in the Caymans, Honduras and Belize. It is pale greyish-white with a black top to the pectoral fin, black edge to the tail, and a dark eye marking that resembles a mask!

1

Juv. 2

3

4

5

6

7

Soapfish

Family: Serranidae

THERE ARE AT LEAST half a dozen species (that we know of) of these grouper-like fish in the Caribbean zone. All are similar in shape, and are widely distributed from Rhode Island out to Bermuda and south to Brazil. They secrete a poisonous mucus that covers the skin, protecting them from predators, particularly when they are lying in wait for prey.

1 GREATER SOAPFISH

Rypticus saponaceus

This species is the largest soapfish, growing to around 30cm (1ft). It is pale grey and mottled brown, with almost a blush behind the large eye. It has rounded, uniform grey fins and its body tapers towards the mouth. It is usually found on its own and is fairly inactive. It tends to rest lying down during the day, often against the corals, and is an active feeder at night on sleeping small fish, hence the larger eyes. It is usually found in depths less than 18m (60ft).

2 FRECKLED SOAPFISH

Rypticus bistrispinus

Fairly rare throughout the range, the Freckled Soapfish is quite small, only growing to around 15cm (6in). It is mainly brown, with cream from the nape to the dorsal fin and all of the underside. A wide brown band tapers through the eye to the snout and spreads over the entire mid length of the body. Usually found resting on the bottom 'playing possum', it is an active predator at night.

Hawkfish

Family: Cirrhitidae

3 REDSPOTTED HAWKFISH

Amblycirrhitus pinos

Vaguely grouper- or hamlet-shaped, this small but colourful fish grows to around 10cm (4in). It is the only member of this diverse family in the Atlantic and Caribbean, and is fairly common throughout the range. It is pinkish brown with a series of wide then narrow brownish bands down the length of the body, and two large black splotches just before the tail and also at the base of the tail. The fins and tail are translucent white, but the head and dorsal fin are covered in red spots. The dorsal fin also has small tassels at the tips of the rays. It is quite wary of divers, preferring deeper recesses in the reef.

Basslets

Family: Grammatidae

BASSLETS ARE A SMALL GROUP of very colourful fish, which are related to groupers. They prefer the deeper reef environs, usually down a cliff wall, and are always around crevices and caverns, where they are often seen swimming upside-down to align with the roof of the cave.

4 FAIRY BASSLET or ROYAL GRAMMA

Gramma loreto

This species has distinctive coloration that makes it a striking sight. The head, elongated pelvic fins and forebody are purple, gradually shading to a brilliant yellow hindbody and tail. It has a distinctive black spot to the rear of the dorsal fin and a black stripe that extends back through the eye. The lips are dull gold, as are a couple of stripes under each eye. It is usually found on the reef walls below 15m (50ft), often in small groups. Occurring from Jacksonville down to Key West, it was first thought to have been introduced to the Caribbean from home aquaria, but it is more likely that its presence is the result of gradual range expansion. It now occurs virtually everywhere in the Caribbean.

1

2

3

4

1 BLACKCAP BASSLET or BLACKCAP GRAMMA

Gramma melacara

Identical in shape to the Fairy Basslet, this solitary species is dark purple with silvery highlights to some of its scales towards the tail. There are some dull golden markings under the eyes, and a black cap which stretches from the upper lip over the head to the dorsal fin. It replaces the Fairy Basslet lower down the reef wall, to around 30m (100ft).

Flounders

Family: Bothidae

FLOUNDERS OR FLATFISH are highly specialised flat-bodied fish which live principally on the seabed. As they mature and transition to lying on their side on the sea floor, one eye moves around the head, so that by adulthood both eyes are side by side on their upper surface.

2 PEACOCK FLOUNDER

Bothus lunatus

The most common flounder throughout the range, this species can grow to around 50cm (1ft 8in) long. It is pale blue with a blue fringe to the fins and tail, and the body is covered in darker blue blotches. It has a long pectoral fin which it raises in alarm. Normally found on the sand flats of the inner reefs and lagoons, it is often partly hidden under the sand for camouflage.

3 EYED FLOUNDER

Bothus ocellatus

A smaller fish at around 15cm (6in), this species is almost circular in shape and is covered in a multitude of different-coloured spots and blotches. When resting on a rough sand or coral rubble seabed, its camouflage is so superb that other flounders may actually rest on top of it!

Blennies

Families: Blennidae, Chaenopsidae

THIS IS A LARGE FAMILY of small fish which live around coastal waters. They like rocky crevices and tend to perch near the entrances of them. Most have fleshy appendages (cirri) above their high-set eyes and, although they resemble gobies, they have shorter, blunt heads, prominent lips, and pectoral fins that function more like feet.

4 DIAMOND BLENNY

Malacoctenus boehlkei

This distinctive fish grows to around 7cm (2¾in). It has a pointed, golden-yellow head which fades to a cream body with grey spots and brownish blotches along the top of the back. The front of the dorsal fin is mottled golden yellow with a very obvious and distinctive dark 'eye'. It has two pairs of cirri above the eyes and on the nape. It lives almost exclusively within the protective grasp of the Giant Anemone (see page 120).

5 SADDLED BLENNY

Malacoctenus triangulatus

This fish has a typical blenny shape, with a pointed head and a uniform body that tapers towards the tail. It is fairly pale, and has five reddish or brown triangular blotches with red spots. There is also a rough band of red spots along the entire length of the body. It has two to four branched cirri above the eyes and on the nape. The small mouth conceals strong teeth for picking out small invertebrates from tight confines. It prefers shallow waters and occurs throughout the entire zone.

1

2

3

4

5

1 ROSY BLENNY

Malacoctenus macropus

This blenny grows to around 5cm (2in). It is pale red, slightly darker on top, with hundreds of tiny red spots over the entire body. The fins are fairly translucent, and speckled with pale yellow. It is rarely found in waters deeper than 10m (33ft) and prefers coral rubble areas and the edges of seagrass beds. It tends to sit still on the bottom, in a 'tripod' stance supported by the ventral fins and tail.

2 DUSKY BLENNY

Malacoctenus gilli

Growing to around 8cm (3¼in), this blenny can be flamboyantly coloured. Its pelvic and anal fins may be red, and the body is covered in dark rosettes. The underside is lighter than the upperside, but the most distinguishing mark is the very obvious blue 'eyespot' on the rear of the dorsal fin, which is usually ringed in cream and then black.

3 GOLDLINE BLENNY

Malacoctenus aurolineatus

This species has the typical blenny shape and grows to around 5cm (2in). It has a wide gold vertical bar midway along the body, and distinct dark markings on the lips, cheeks and underneath the eyes. It is usually found in waters less than 12m (39ft) deep.

4 HAIRY BLENNY

Labrisomus nuchipinnis

Another shallow-water blenny, this species is somewhat elongated, to a maximum of 23cm (9in). It has spots over all of the body and fins, and is slightly darker on the top half of the body. It has large eyes with radiating darker lines around them, blending into the rest of the coloration, and also has a dark spot just behind the gill covers.

5 SEAWEED BLENNY

Parablennius marmoreus

A classic blenny in shape with a long tapered body and blunt head, this species has large cirri above the eyes with three or four branches. It has bold chequered markings along the entire body. It occurs in mixed habitats of fan corals, algae and coral rubble, and is quite shy and will hide if you get too close. However, curiosity soon brings it back out again.

6 MOLLY MILLER

Scartella cristata

The Molly Miller can grow to around 10cm (4in). It has a rounded head and large fleshy lips. It is characterised by its two sets of fringes, one just above the eyes (resembling eyelashes), and a second set further up on the nape. It enjoys rockpools and shallow water of less than 6m (20ft) and quite often occurs in rougher water along the shore.

7 REDLIP BLENNY

Ophioblennius atlanticus

A common blenny found throughout the entire region, this species has a very blunt head (the front is almost vertical). It has obvious red lips and is either entirely chestnut-coloured, or has a chestnut head that fades to grey and pale brown towards the tail. It perches on coral heads, sponges and coral boulders and is quite wary of divers, although it makes a great photographic subject if you can get close enough!

1

2

3

4

5

6

7

1 SAILFIN BLENNY

Emblemaria pandionis

Curious and comical in its behaviour, this small blenny can grow to around 5cm (2in). It is usually a uniform dark brown, with small indistinct stripes around the head and a couple of cirri above the eyes. It lives in old, empty worm casts and will vigorously protect its territory by emerging almost entirely out of its burrow, showing a huge sail-like dorsal fin, and waving its long pelvic fins.

2 RED BANNER BLENNY

Emblemariopsis ramirezi

Fairly similar to the Sailfin Blenny, this species has an identical lifestyle. It is distinguished by the red banner to the top of the forward part of the dorsal fin, underlined by a white stripe with black below. It is rarely reported from the Caribbean, but this individual was in the Cayman Islands.

3 DARKHEADED BLENNY

***Emblemariopsis* sp.**

A number of very similarly coloured and shaped species, often referred to as 'glass blennies', these fish are almost impossible to tell apart except for a few distinguishing spots or markings about the head. The body is virtually transparent, the spinal cord clearly visible and marked with red dots and dashes. Quite small at around 2.5cm (1in), they are most often found hiding in a former tube-worm hole, but they also rest on the tops of boulder corals and sit amongst the coral polyps. However, they are almost impossible to see.

4 SPINYHEAD BLENNY

Acanthemblemaria spinosa

This species and the two that follow are also very similar, but with more obvious markings. The Spinyhead Blenny grows to around 2.5cm (1in), and its almost white or greyish-blue forehead and snout are liberally covered in tiny spines. The tapering body is usually dull dark grey, and the eye large with a pale yellow ring around the pupil. It inhabits old tube-worm holes or pores in sponges.

5 ROUGHHEAD BLENNY

Acanthemblemaria aspera

Similar in size to the Spinyhead Blenny, the female of this species is usually golden yellow. The male is grey and brown with plentiful dots, spots and dashes, and an ocellated spot at the front of the dorsal fin. It has quite long cirri above the eyes as well as a small, often orange-red, clump between the upper lip and eyes.

6 SECRETARY BLENNY

Acanthemblemaria maria

This is the most common and the largest of the three *Acanthemblemaria* species described here, growing to around 5cm (2in). It has a yellowish eye and a series of small spikes over the front of the pale green head. There is a very obvious white rounded streak below the eye that extends to the gill covers. The body is dark with darker vertical bars and spots. It occurs throughout the entire region, except for Florida and north-east to Bermuda. It lives in old tube-worm holes, especially in brain corals.

7 ARROW BLENNY

Lucayablennius zingaro

The Arrow Blenny grows to around 5cm (2in). Its body is arrow-shaped – long and thin with a long narrow snout and tiny mouth. It is reddish with a cream stripe from the upper lip to the start of the dorsal fin. It has a golden tail with black spots, which it keeps 'cocked' at all time. It tends to hover over the reef, but should danger threaten, it flicks its tail and disappears to another part of the reef in the blink of an eye. It is usually found below the barrier reef crest around 12–30m (40–100ft). It likes underhanging corals and sponges.

1

2

3

4

5

6

7

1 ROUGHHEAD TRIPLEFIN

Enneanectes boehlki

Growing to around 4cm (1½in), this triplefin's first dorsal fin is shorter than the others. It has five obvious dark bars down an otherwise red body, and its pectoral fins have a couple of rows of bright yellow spots. The head may also have some pale blue or cream spots.

2 LOFTY TRIPLEFIN

Enneanectes altivelis

This species grows to around 4cm (1½in). It has a long, pointed, wedge-shaped snout. The first dorsal fin is quite tall and is usually cream or white. It tends to live amongst various sponges.

3 YELLOWFACE PIKEBLENNY

Chaenopsis limbaughi

This species is similar to the Bluethroat Pikeblenny, but grows to only 10cm (4in) and its throat bears yellow rather than blue stripes. Uncommon in the range, it prefers deeper water to 18m (60ft), and hides in old worm tubes.

4 BLUETHROAT PIKEBLENNY

Chaenopsis ocellata

This very distinctive fish grows to almost 15cm (6in). It has a long pointed snout, a yellow face and large lower jaw which opens up to show blue stripes in the throat folds. It is only found in shallow water of less than 3m (10ft), but is shy and rarely seen. It lives tail-first in old worm tubes.

Gobies

Family: Gobiidae

GOBIES HAVE TWO dorsal fins and hold themselves quite rigidly when at rest. The pelvic fins are generally modified into a sucker disc, which keeps the fish securely in place as it rests on an underhanging rock ledge. Most gobies are bottom-dwellers and, although they live in similar habitats, are a diverse group in both appearance and habits. There are a number of very similar species and it takes patience to get close enough to be able to properly identify them.

5 NEON GOBY

Gobiosoma oceanops

The Neon Goby's body is very dark blue, with a neon blue stripe that starts just before the eye, passes over the eyebrow and continues to the tip of the tail. This is a cleaner fish and will set up cleaning stations on the reef anywhere between 3–30m (10–100ft). It is more commonly found in the central Caribbean and from Florida to Honduras, but rarely in the Bahamas.

6 CLEANING GOBY

Gobiosoma genie

This is probably the most common cleaning goby and can be found on almost all of the reefs, except Florida and the eastern seaboard. The body is uniformly dark with a brilliant yellow 'V' on the snout that runs over the eyebrow, fading to a pale gold along the body. It has a pale underside, usually with a pinkish blush. It congregates in quite large numbers to clean the local reef residents of parasites.

1

2

3

4

5

6

1 SHARKNOSE GOBY

Gobiosoma evelynae

This species grows to around 4cm (1½in). It is very similar to the Cleaning Goby, but the yellow 'V' on the snout fades to iridescent blue along the length of the body. The underside is pale with a pinkish blush. It is completely unafraid of divers; if you approach slowly and extend your fingers, it will explore in search of titbits. It prefers the lower reef.

2 YELLOWLINE GOBY

Gobiosoma horsti

Growing to around 5cm (2in), this species is slightly larger than those previously described. It has a pair of bright yellow lines that start over the eyes and taper away to the end of the tail; it has no other markings. It occurs most often amongst sponges on the lower reef, around 15–18m (50–60ft).

3 SPOTLIGHT GOBY

Gobiosoma louisae

The Spotlight Goby is similar in every way to the Yellowline Goby, except for a small spot of yellow on the snout between the eyes. The body is almost black, with a prominent yellow line that runs from the eyebrows to the tip of the tail.

4 YELLOWPROW GOBY

Gobiosoma xanthiprora

Very similar to the Spotlight Goby, but smaller in size at around 2.5cm (1in), this species has a longer dash on the snout instead of a dot. It has a dark upper body and paler underside with a pink blush. A strong yellow stripe runs from above the eye to the tip of the tail.

5 YELLOWNOSE GOBY

Gobiosoma randalli

This species resembles the Yellowprow Goby and also has a yellow snout, but the colour here is much more golden and stretches from the mouth, over the top of the eye to the tail. Its mouth is pinkish and it displays bright blue stripes between the golden lines.

6 ORANGESIDED GOBY

Gobiosoma dilepsis

One of the more colourful gobies, this species grows to a maximum of 2.5cm (1in), but is usually much smaller. It is pale with red markings on the head and between the eyes, and large squarish red blotches edged in black along the length of the body. It is usually found on coral heads in depths over 9m (30ft).

7 MASKED or GLASS GOBY

Coryphopterus personatus

The Masked Goby is found from Bermuda to the rest of the Gulf of Mexico. It can reach a maximum size of 4cm (1½in), and is mostly transparent with a yellowish head, dark eyes and a row of six white dashes along the middle of the body, starting on the upper eye. It is often found in small groups. The very similar *C. hyalinus* has more yellow, and is much rarer throughout the southern Caribbean.

8 PEPPERMINT GOBY

Coryphopterus lipernes

A strikingly coloured small goby, this species grows to around 4cm (1½in) and is a bright, translucent yellow with an electric blue eye and snout. These lines may extend behind the eye, but the body is predominantly yellow. It prefers to sit on coral heads and sponges and is unafraid of divers.

1
2
3
4
5
6
7
8

1 BRIDLED GOBY

Coryphopterus glaucofraenum

This goby can grow to around 7.5cm (3in), its body gradually changing shape with age. An electric pale blue stripe extends from behind the mouth and underneath the eye. The body varies from almost transparent when it is young to dark with many blue dots, and a pale row of dashes along the midline of the body.

2 PALLID GOBY

Coryphopterus eidolon

Usually sand-dwelling or frequenting small patches of coral rubble, this goby grows to around 5cm (2in) and has a slightly longer snout than most others. It is pale brown and cream with a dark band between the eyes and a yellowish band that extends back behind the eyes.

3 COLON GOBY

Coryphopterus dicrus

Quite translucent with brown and white markings, this goby can reach 5cm (2in). The base of the tail has two spots, almost connected. It can be found from southern Florida to Brazil and is relatively unafraid of divers.

4 GOLDSPOT GOBY

Gnatholepis thompsoni

Growing around 5cm (2in) long, this pale goby has a very obvious vertical black line down through the eye from the top of the head and a less prominent gold spot located just above the pectoral fin. The body may also have squarish blotches. It is fairly common in all areas.

5 DASH GOBY

Ctenogobius saepepallens

The Dash Goby can grow to around 5cm (2in), but is usually much smaller. It is quite pale with a series of dark dashes along the lower half of the body, and a vertical bar down through the eye. It is often seen in mating pairs.

6 ORANGE SPOTTED SHRIMPGOBY

Nes longus

The only known member of its genus, this species is found throughout the Caribbean but is very rare, probably due to its very reclusive behaviour. It is extremely wary of everything, especially underwater photographers! It lives in association with the Burrowing Partner Shrimp (*Alpheus floridanus*, see page 158), standing guard outside a shared burrow, which is excavated by the shrimp. When danger threatens, both retreat into the burrow and it takes some considerable time before the goby will emerge and signal to the shrimp that all is well with a flick of its tail.

7 GREENBANDED GOBY

Gobiosoma multifasciatum

A small goby, rare but found in the central Caribbean around the Cayman Islands. It grows to around 5cm (2in). Its most striking feature is the broad reddish stripes that extend from the snout through the eyes to behind the gill covers. It also has very obvious neon green bands down the length of the body.

1

2

3

4

5

7

6

Dragonet

Family: Callionymidae

1 LANCER DRAGONET

Paradiplogrammus bairdi

A small fish, growing to around 10cm (4in), the Lancer Dragonet is generally pale and rather drab with a mottled coloration. During mating, the male develops vivid blue streaks to the head sides, and a yellow snout. This species is uncommon throughout the range.

Clingfish

Family: Gobiesocidae

2 STIPPLED CLINGFISH

Gobiesox punctulatus

Slightly larger than the Red Clingfish at 5cm (2in), this clingfish has tiny black spots all over the body and may have different-coloured bands down its length. Being so small, it is rarely seen, and it generally hides away under stones during the day.

3 RED CLINGFISH

Arcos rubiginosus

This species is around 3cm (1¼in) long, and is reddish brown with paler radiating lines below the large, red eye and numerous spots and freckles over the body. It has adapted pelvic fins which allow it to grip onto a surface. It is active at night and is usually found on the soft coral sea rods and sea fans; it is regarded as uncommon everywhere in the range.

Jawfish

Family: Opisthognathidae

THESE ARE SMALL, elongated fish with a continuous long dorsal fin. They rarely emerge from their seabed holes, which are usually on fine sand or coral rubble. They will venture out to feed on passing zooplankton but stay about a tail's length away from safety, retreating tail-first into their burrow if you get too close. During the mating season, the males brood the egg mass in their mouths for protection.

4 YELLOWHEAD JAWFISH

Opstognathus aurifrons

The most colourful of the Caribbean jawfish, this species can grow to around 10cm (4in), but is usually smaller. It has a bright yellow head blending into an electric blue tail; the other fins are also bright blue. It feeds on plankton and may be seen hovering above the seabed. Like most jawfish it is a 'mouthbrooder'; the male incubates the eggs in its mouth.

5 DUSKY JAWFISH

Opistognathus whitehursti

A similar size to the Yellowhead Jawfish, this species is rarer and more timid – usually all you will see is a reddish head protruding from the burrow. It has a light stripe running the length of the body, underneath the dorsal fin.

6 BANDED JAWFISH

Opistognathus macrognathus

The largest of the Caribbean jawfish, the Banded Jawfish grows to around 20cm (8in). It is a rather uniform pale greyish tan, with a dark spot at the front of the dorsal fin and a series of indistinct blotches along the length of the body. It will usually tolerate a slow approach. It occurs from Florida through the Bahamas, and into the central Caribbean.

1
2
3
4
5
6

Squirrelfish and Bigeyes

Families: Holocentridae, Priacanthidae

THERE ARE AROUND A DOZEN representatives of this group in the Caribbean and western Atlantic. They are mainly red, with large eyes. The rear of the dorsal fin is long and they all have forked tails, except for the Bigeye and Glasseye which have flat tails. All of the species are similar, and examining the small markings on individuals is necessary for identification. They are nocturnal feeders, hunting invertebrates amidst the coral rubble, and are themselves favourite prey for groupers.

1 SQUIRRELFISH

Holocentrus adscensionis

This is the most common of the squirrelfish. It grows to around 30cm (1ft) but can be much larger. It is reddish with silver markings and a pronounced bar down the body at the rear of the dorsal fin. The tail and pelvic and anal fins are white. The dorsal fin is tinged yellow and is often erect with a spiky top. As with other squirrelfish, it has a large dark eye, ringed in red. The underside is silvery white. It often lives in small groups and occurs throughout the entire range.

2 LONGSPINE SQUIRRELFISH

Holocentrus rufus

Very similar to Squirrelfish, this species is usually smaller at around 25cm (10in). It has more extensive red coloration, through to the tail, with more obvious lateral stripes along the length of the body. The dorsal fin has a little white flag on the tips.

3 DUSKY SQUIRRELFISH

Sargocentron vexillarium

This is a smaller squirrelfish, reaching around 12.5cm (5in). It is bronzy red to brown, with silvery dotted longitudinal stripes along the length of the body. The dorsal fin has dark spots at the front. It is fairly common throughout the entire region.

4 LONGJAW SQUIRRELFISH

Neoniphon marianus

This species grows to around 15cm (6in), and has a longer snout than other squirrelfish. It is silvery red with golden stripes along the edge of the body, with a silvery white underside to the jaw. The dorsal fin has white tips to the spines, and white 'flags' halfway up. The anal fin has a very long spine.

5 ATLANTIC BIGEYE

Priacanthus arenatus

Bigeyes as a group are found in all tropical waters and are represented in the western Atlantic and Caribbean by a handful of species. This species grows to around 30cm (1ft). It is dull red in colour and is common all over the range, from Bermuda south to Argentina.

6 BLACKBAR SOLDIERFISH

Myripristis jacobus

The Blackbar Soldierfish grows to around 20cm (8in). It is red with a very obvious black bar just behind the head. The fins are edged silvery white. It is often found in medium to large schools and is frequently associated with shipwrecks and deep reef caverns. One of the more common species in the group, it occurs throughout the entire range from Bermuda down the Carolinas and over all of the western Atlantic and Caribbean.

7 GLASSEYE SNAPPER

Heteropriacanthus cruenatus

Often known just as Glasseye, this species is almost identical to the Atlantic Bigeye, but has silver bars down the back and a metallic appearance. It can grow to 30cm (1ft). It is frequently seen in fairly large schools, often in a vertical formation.

1

2

3

4

5

6

7

Cardinalfish

Family: Apogonidae

SMALL FISH WITH LARGE EYES, these species have a slightly elongated snout and large mouth, and most have bands and spots towards the tail. They are nocturnal and deep shade-dwelling hunters, and feed on plankton. Like the jawfish (see page 70) the male broods the egg mass in his mouth until they hatch. Cardinalfish live on the lower reef.

1 BARRED CARDINALFISH

Apogon binotatus

One of the more common species found throughout the Caribbean and Gulf of Mexico, the Barred Cardinalfish grows to around 5cm (2in). It is pale orange or red with two thin bars down the body, the first just behind the second dorsal fin and the second just before the base of the tail. It can be seen in small groups, mostly on night dives as it hides during the day.

2 PALE CARDINALFISH

Apogon planifrons

This species is almost identical to the Barred Cardinalfish, with similar markings, but is a pale silvery pink or pale orange colour.

3 CARIBBEAN BANDED CARDINALFISH

Apogon townsendi

This species reaches 5cm (2in). It is often seen alone, but groups together in the evenings, often at the mouths of caves and caverns, before venturing out at night to feed.

4 MIMIC CARDINALFISH

Apogon phenax

Very like the previous two species in markings and habits, the Mimic Cardinalfish is slightly smaller and a much darker red in colour.

5 WHITESTAR CARDINALFISH

Apogon lachneri

Another small cardinalfish, this species grows to around 5cm (2in) or slightly larger. It is more rotund than the others, has a large black eye for feeding at night, and has two very prominent dorsal fins. It prefers good clean water and open reefs, and is distinguished by the black and white spot at the rear base of the second dorsal fin.

6 DUSKY CARDINALFISH

Phaeoptyx pigmentaria

This cardinalfish grows to around 7cm (2¾in). It is pale red with fine dark spots all over the body from the snout to the dark bar at the base of the tail. All of the fins are clear and unmarked, and there is a smudged bar below the eye.

7 BRIDLE CARDINALFISH

Apogon aurolineatus

Only growing to around 5cm (2in), this species has no additional markings on the body, but the tail may be a light gold in colour, with the rest of the body rusty. It lives close to the tentacles of anemones for extra protection, and is always in small groups.

1

2

3

4

5

6

7

Hatchetfish

Family: Pempheridae

1 GLASSY SWEEPER

Pempheris schomburgki

Vaguely hatchet-shaped, this small fish can grow to around 15cm (6in), but is usually much smaller than this. It is copper-coloured, with a dark band just at the base of the anal fin. It inhabits deeper recesses, caverns, and deeply shaded areas inside wrecks.

2 Silversides

Families: Atherinidae, Clupeidae, Engraulidae

Silversides are juvenile fish from about ten different species, which all group together. Most are herrings, anchovies and boga, but it is impossible to make a correct identification, as the tightly packed polarised schools move as one, staying out of the way of divers as they drift and move through the coral canyons, under jetties and inside wrecks.

Scorpionfish

Family: Scorpaenidae

SCORPIONFISH ARE bottom-dwelling stealth hunters with large heads that usually bear fleshy appendages. They have mottled coloration, providing camouflage as they lie in wait for prey to approach. When prey comes close enough they lunge, opening their wide mouths at the same time and sucking in water and prey together. They have venomous spines along the dorsal fin, but their toxin (although very painful) is not lethal, unlike the Stonefish that occurs in other tropical waters. There are no Stonefish in the Caribbean, despite what dive guides may tell you!

3 PLUMED SCORPIONFISH

Scorpaena grandicornis

The Plumed Scorpionfish grows to a maximum of 17cm (7in). It has a pair of large fleshy cirri or fleshy appendages over its eyes. There may also be numerous cirri on the cheeks, snout and under the lower lip. It is multicoloured, its patterning providing superb camouflage that renders it almost invisible. It lives amongst seagrasses, algae and coral rubble, and occurs from Bermuda to Brazil and throughout the region, but it is difficult to find.

4 SPOTTED SCORPIONFISH

Scorpaena plumieri

This is the largest scorpionfish, growing to around 50cm (1ft 8in). It is otherwise very similar to the Plumed Scorpionfish, and often has plumed cirri over the eyes as well as various skin flaps under the chin. The overall colour is beige, brown and mottled tan except for the pectoral fins, which are a vivid red, purple and black with white spots. These colours are only seen when the fish is ‘in flight’, gliding over the seabed.

5 REEF SCORPIONFISH

Scorpaenodes caribbeus

The smallest of the region’s scorpionfish, this species grows to a maximum of 10cm (4in), but is usually much smaller. It is multicoloured and superbly camouflaged. It is difficult to find as it is very shy and hides amongst the soft corals and a variety of sponges, and often swims upside-down in small caverns and overhangs.

Lionfish

Family: Scorpaenidae

THE COMMON NAME lionfish refers to quite a number of species worldwide, including two invasive species that are now found throughout the Caribbean, Gulf of Mexico and even as far as Bermuda. First discovered in the 1980s, they were thought to be accidental aquarium releases after a hurricane. Conditions appear perfect for them as they have few natural predators, except for some of the larger groupers. The venom in their spines is extremely painful and medical attention will be required – I know from personal experience!

1 RED LIONFISH

Pterois volitans

The most common lionfish found in the region, this species grows to 40cm (1ft 4in) long and has a distinctive brown and cream striped body and fins. Using unique swim bladders, it can position itself anywhere in the water column, aligning itself with various prey species of fish or invertebrate. It has a wide, extending mouth and is a voracious hunter on all types of reef and wreck. It is found in all depths, at least as far as 60m (200ft).

2 COMMON LIONFISH or DEVIL FIREFISH

Pterois miles

Almost identical to the previous species, although much rarer in this region, this lionfish is equally at home in its new environment and both species are regarded as a very real threat to the reefs and reef fish on the Caribbean zone. It has small white spots, a variable number of spines, and the fins are often black and radiate out from the body. When it sits still with its fins extended, it very closely resembles the Long-spined Urchin (see page 166).

Frogfish

Family: Antennariidae

FROGFISH ARE FOUND in all tropical waters and a number of species have been recorded from the Caribbean and Gulf of Mexico, although they are considered rare throughout the entire region. Vaguely resembling scorpionfish, they rely on their camouflage and will not move unless disturbed. It is usually only their eyes or upturned mouth that are noticeable. Rather than swimming, they hop or waddle on their adapted pectoral fins. They hunt with the aid of a lure which they extend over the mouth and wriggle about to attract prey. They lunge at their prey, opening their large mouths wide and sucking in the hapless victim.

3 SARGASSUMFISH

Histrio histrio

Resembling Sargassum Algae (see page 210), this frogfish has a huge distribution and is both subtropical and circumtropical. It can grow to around 20cm (8in) but is usually much smaller. Coloured green, brown and cream, with many small protuberances and fleshy tabs all over the body, it blends in very well with the seaweed. I first encountered this fish in Bermuda many years ago, but have not seen one since!

4 LONGLURE FROGFISH

Antennarius multiocellatus

This is the most common of the Caribbean frogfish, and grows to a maximum of 20cm (8in). It is highly variable; I have seen red, yellow, pink and multicoloured versions. It tends to rest on sponges that match its own colour, and its numerous spots mimic the openings on the sponge. Whilst it can be locally common, it is rarely seen due to its excellent camouflage. I have encountered this species in Bonaire, Cuba, Honduras, St. Lucia and the Cayman Islands.

1

2

3

4

Batfish, Toadfish, Searobin and Gurnard

Families: Ogcocephalidae, Batrachoididae, Triglidae, Dactylopteridae

1 POLKA-DOT BATFISH

Ogcocephalus radiatus

This is a curious, bottom-dwelling fish that grows to around 38cm (1ft 3in). It is multicoloured and speckled with small spots, providing camouflage. The pelvic fins are located quite far forward, and are flared and have arm-like stalks. It uses these fins to waddle around the seabed until it finds a suitable spot to lie in wait and use its lure to attract unwary shrimps and juvenile fish. The lure is a retractable projection found between the eyes, which the fish wiggles around; the lure also produces a chemical that also helps attract prey.

2 SPLENDID TOADFISH

Sanopus splendidus

This striking species has bright yellow fins and a black and white or grey-striped head. As a bottom-dweller, its eyes are on top of its broad, flattened head, and it has large barbels around and under its mouth. Associated with the reefs off Cozumel and Cancún in the Yucatán, it is fairly common here and often found by its croaking noise, which gives it its name. It is usually found at the edge of the reef, and always under an overhang.

3 LARGE EYE TOADFISH

Batrachoides gilberti

Quite a large fish, this species can grow to about 30cm (1ft). It has a broad flattened head and is usually coloured in various shades of brown, sometimes with oblique stripes towards the tail. It has a large number of branched barbels to the side and underneath the mouth. It prefers a soft substrate, often in brackish waters.

4 OYSTER TOADFISH

Opsanus tau

This none-too-handsome fish goes by a number of unflattering names, including 'ugly toad' and 'bar dog'! The largest of the toadfish, it can grow to around 43cm (1ft 5in). It buries itself in soft substrate up to its eyes, and lies in wait for unwary prey, mainly worms, shrimps and crabs. It is found mainly on the eastern Gulf of Mexico and upwards to Maine. Weirdly, a NASA project sent this species into space to see what the effects of microgravity would be on organ development; they discovered that there was little difference!

5 LEOPARD SEAROBIN

Prionotus scitulus

Very similar in shape to the European gurnards, with their shovel-like heads and reddish coloration, this species wanders over the seabed with its adapted pelvic fin rays, three of which are very versatile and which it uses to forage for food. It has eight dorsal spines and four dark vertical blotched bars on the side of the body. The eyes are positioned on top of the head, indicating that it does spend some time hiding under the substrate. It has obvious tan and brown and bluish spots all over its huge and extendable pectoral fins which allows it to 'fly' over the seabed.

6 FLYING GURNARD

Dactylopterus volitans

The Flying Gurnard is known from most tropical waters and is a bottom-dwelling fish found on both sides of the Atlantic. A very similar species occurs in the Indo-Pacific. It can grow up to around 50cm (1ft 8in) and weigh as much as 1.8kg (4.0lb). Its blueish pectoral fins can extend like huge wings, enabling it to 'fly' over the seabed. Usually found in pairs, it occurs regularly on the sand planes before a reef wall drops off into the depths, as well as inside lagoons and on the shelf between the fringing and barrier reefs.

1
2
3
4
5
6

Lizardfish

Family: Synodontidae

LIZARDFISH HAVE LONG cylindrical bodies, with a wide, slightly upturned head and a wide mouth jammed full of teeth. They have two dorsal fins: one very large, which they fold back against the body, the other tiny. They will hide under the sand as well as perch on corals and sponges, and are bottom-dwellers.

1 SAND DIVER

Synodus intermedius

The Sand Diver grows to around 40cm (1ft 4in). The markings along the length of the body can be quite pale, with more obvious lines underneath the dorsal fins.

2 BLUESTRIPED or ATLANTIC LIZARDFISH

Synodus saurus

This species is almost identical to the Sand Diver in size and colour. It is light tan and mottled brown, which helps its camouflage when it dives for cover into the sand. It is common in shallow sandy areas where it hides under the sand. Quite often, all that is seen are the eyes and upturned mouth.

Seahorse and Pipefish

Family: Syngnathidae

THESE CHARISMATIC FISH are strange and exotic, and are loved by just about everyone thanks to their comical appearance. Seahorses are named for their horse-like cocked head. They have an upright posture when swimming, and a stout body that tapers to a prehensile tail, with which they anchor themselves around sponges, algae and other seahorses. The fertilised eggs are passed to the male who gestates them in a specially adapted brood pouch, until they hatch. Pipefish are long, linear and live on or near the seabed. Their heads blend into their eel-like bodies.

3 LONGSNOUT or SLENDER SEAHORSE

Hippocampus reidi

Regarded as fairly rare throughout the range, this seahorse can be locally common and is often found under docks and piers. It varies in colour, though is most often dull dark grey, brown or black. It is quite shy and will just lie about on the seabed pretending to be a bit of algae.

4 PIPEHORSE

Acentronura dendritica

Bridging the gap between the seahorses and pipefish, the Pipehorse looks like a very slender and straightened-out seahorse, and can grip onto seaweeds with its tail. Its fins look like little tufts of weed. It has numerous other fleshy tabs, helping to provide great camouflage.

5 HARLEQUIN PIPEFISH

Micrognathus ensenadae

This is a colourful, long, eel-like fish, growing to around 20cm (8in). It is banded with brown and yellow stripes. It has prominent eyes, a long snout and a small mouth for taking in tiny invertebrates. It lives amongst algae and seagrasses, but is equally at home on the reef and is an active predator at night. It is found from Bermuda to Brazil and throughout the Caribbean area.

6 SHORTFIN PIPEFISH

Cosmocampus elucens

Growing to around 15cm (6in), this pipefish is reddish brown with numerous bands of other colours along its entire length. The eyes (usually red) are mid-head, and it has a long snout and slightly upturned mouth. Most commonly found in shallower water and almost always on the seabed, it hunts through the coral rubble and seagrasses for small invertebrates.

1
2
3
4
5
6

1 BANDED PIPEFISH

Micrognathus crinitus

Growing to around 15cm (6in), this species has a short snout and is usually pale golden with wide indistinct bands down the length of the body. It prefers algae and seagrass beds, in waters usually shallower than 12m (39ft), and is widespread throughout the range.

2 PELAGIC or SARGASSUM PIPEFISH

Sygnathus pelagicus

This is a pelagic pipefish, found travelling with floating Sargassum Algae. It can grow to around 20cm (8in) and is coloured in various shades of cream, brown, beige and white. It occurs on both sides of the Atlantic and the Mediterranean, and even the western Indo-Pacific region.

Cornetfish, Trumpetfish and Needlefish

Families: Fistularidae, Aulostomidae, Belonidae

3 RED CORNETFISH or ROUGH FLUTEMOUTH

Fistularia petimba

This fish is eel-like with a very long thin body and long tail filament. It can grow to 1.8m (5ft 11in), although rarely exceeds 1m (3ft 3in). It is pale red, often with dots and blotches all over the body. The long snout's lower jaw is longer than the upper. It tends to hide amongst sea rods (*Plexaura*), where it is well camouflaged. Although rare, it is found throughout the western Atlantic, greater Caribbean and all tropical oceans.

4 TRUMPETFISH

Aulostomus maculatus

The Trumpetfish grows to around 90cm (3ft). It has a trumpet-like mouth and is reddish brown in colour, but may also be yellow as it can change colour to suit its surroundings. It is common from Bermuda to Brazil, Florida, the Bahamas and all through the Caribbean and Gulf of Mexico.

5 KEELTAIL NEEDLEFISH

Platybelone argalus

This is a long and slender fish that grows to around 60cm (24in). The beak is elongated and slender with the upper jaw slightly shorter than the lower, and there is a wide and flattened keel towards the base of the tail. It often forms small hunting groups at dusk and night, always near the surface, feeding on plankton.

Tilefish

Family: Malacanthidae

6 SAND TILEFISH

Malacanthus plumieri

This fish has a silvery white body and grows to around 60cm (2ft). It lives in sand and loose coral rubble areas, often in spur and groove reef structures, and is usually seen hovering over the sand. It constructs 'nests' by moving small stones, shells and coral rubble into a loose mound with an entrance and exit, where it dives for cover if danger threatens.

1

2

3

4

5

6

Jacks

Families: Carrangidae, Scombridae

JACKS ARE HIGHLY SPECIALISED, powerful swimming fish, more at home in the open ocean than near the shore. All have very streamlined bodies with a narrow tail base and forked tail. The pectoral fins are scythe-shaped. Most species resemble members of the tuna family, although are much slimmer. Most jacks are fish-hunters and are themselves an important source of food for human consumption.

1 BAR JACK

Caranx ruber

This is one of the more common jacks to be found throughout the region. It can grow to a maximum of around 60cm (2ft). Its body sides have a dark bar which extends backwards under the dorsal fin and terminates at the top of the lower tail fin. Bar Jacks can change colour from their typical metallic silvery blue to almost black, and often when encountering two fish together, one will be black (as if it were the shadow of the other!). The Bar Jack is an opportunistic feeder and often associates with stingrays and other large fish.

2 BLUE RUNNER

Caranx crysos

Similar to the Bar Jack, this species can be longer at almost 90cm (3ft). It lacks the dark band of the Bar Jack, but has dark tips to its thin forked tail. It also has a darkish blotch near the end of the upper gill cover. Known more for being an offshore pelagic fish, it can be found in large schools and appears interested in divers.

3 YELLOW JACK

Caranx bartholomaei

Another large jack, similar in size and shape to the previous two, this species has a yellowish tinge and yellow spots on the otherwise silvery body, and a yellow tail and fins. It has quite a high back. Often seen in pairs, it also enjoys a solitary life hunting along the reef crest.

4 HORSE-EYE JACK

Caranx latus

This jack grows to around 1m (3ft 3in). It has very large eyes, a dark edge to the dorsal fin, sometimes a dark blotch at the base of the pectoral fin, and a line of black scutes towards the yellow tail. Often found in quite large schools fairly inshore, at the reef crest or sand plane and certainly around some shipwrecks, it is fairly common throughout the western Atlantic, Gulf of Mexico and Caribbean.

5 CREVALLE JACK

Caranx hippos

This is a large jack, growing over 1m (3ft 3in). It is elongated with a large round head, and has a dark blotch at the base of the pectoral fin and a black mark near the top of the gill cover. It leads a pelagic lifestyle and rarely swims over reefs, instead patrolling the outer reef edge, often in fairly large schools. It is quite common and wide-ranging, from Nova Scotia south to Uruguay.

6 BLACK JACK

Caranx lugubris

The Black Jack grows to 90cm (3ft). It has a high forehead and is usually dark brown or grey. The dorsal fin and outer edge of the tail are black, and the body has a prominent line of black scutes. It is often seen in pairs and, as a deep-water fish, is rarely found over reefs, preferring the edge of a submarine wall. Whilst it is considered uncommon, its distribution is circumtropical.

1

2

3

4

5

6

1 LOOKDOWN

Selene vomer

This strikingly shaped fish has a very high forehead, giving the impression that it is looking down on the rest of us! It is about 30cm (1ft) long and is bright metallic silver, often tinged gold or blue. It can be found in small schools and occurs on both sides of the Atlantic, as well as throughout the Caribbean zone.

2 PERMIT

Trachinotus falcatus

Very similar in shape to the Snubnose Pompano (*Trachinotus blochii*) found in other tropical waters, the Permit can easily reach 1m (3ft 3in) in length. It has a large rounded forehead and its dorsal and anal fins are positioned far back on the body. The fronts of these fins are elongated. The pectoral fin is black, as is a large irregular blotch along the flanks, leading to a black-edged, forked tail. It is usually found on its own.

3 PALOMETA

Trachinotus goodei

This jack grows to around 50cm (1ft 8in). It has extended dorsal and anal fins and a deeply forked tail, all fringed in black. There are four darkish bars down the length of the body. Often found in small groups in fairly shallow water, it is considered to be quite rare in the north-western Caribbean.

4 COTTONMOUTH JACK

Uraspis secunda

Growing to just over 30cm (1ft), this small, rounded jack lacks obvious markings and has low and elongated dorsal and anal fins. It is pelagic and often forms large schools.

5 GREAT AMBERJACK

Seriola dumerili

As the name suggests, this jack is large, reaching around 1.8m (5ft 11in). It has a light but distinct diagonal stripe which runs from the upper lip, through the eye and terminates on the back of the head.

6 ALMACO JACK

Seriola rivoliana

Half the length of the Great Amberjack, this species has a strong diagonal stripe from the lip to the nape. Although it is considered to be an open-water pelagic fish, It can be encountered in small groups close to the reef.

7 MACKEREL SCAD

Decapterus macarellus

This is a slender jack, growing to around 30cm (1ft). It has an obvious dark blotch at the base of the pectoral fin, and is otherwise neutral silver colour. It tends to form large schools, and often mingles with other similarly shaped fish. It enjoys life around the reefs and has a circumtropical and subtropical range.

8 BIGEYE SCAD

Selar crumenopthalmus

Growing to a maximum of around 60cm (2ft), this fish's silvery body is highly reflective. It is elongated with a deeply forked tail. It is often found in simply staggering numbers; my experience with a vast school was under the Salt Pier in Bonaire, and the fish literally blocked out the light above.

1

2

3

4

5

6

7

8

1 BOGA

Inermia vittata

Growing to around 20cm (8in), this species has a silvery body, more blue on the upperside, with thin, dark lateral stripes along the body and a yellowish snout. It is often found in large schools, sweeping along the edge of the reef walls or in open water with schools of other reef fish, usually below 15m (50ft). Although it is uncommon, it can be found throughout the range.

2 PILOT FISH

Naucrates ductor

Growing to around 70cm (2ft 4in), this fish is silvery grey with dark blue/black transverse lines all along the body. It is usually seen accompanying sharks, large jacks, tuna and turtles, cleaning off their parasites and eating any left-over food scraps.

Barracuda

Family: Sphyraenidae

3 GREAT BARRACUDA

Sphyraena barracuda

The Great Barracuda can grow to over 1.75m (5ft) long, with large specimens weighing around 40kg (100lbs). Tubular in shape, it has a ferocious look and a mouthful of pointed, sharp teeth, set in two rows. It is a metallic silver colour, sometimes with darker spots and blotches. It is extremely fast when hunting, reaching speeds of more than 56km/h (35mph), and often in small packs when younger, changing to a more solitary life in adulthood. It occurs throughout the world.

Tarpon

Family: Elopidae

4 TARPON

Megalops antlanticus

This is a large, elongated silver fish that can grow well over 2m (6ft 6in). It has a pointed snout and huge upturned mouth. It prefers inshore waters where it hunts in small groups at all times of day or night, principally preying on the silverside schools that inhabit the recesses of wrecks and the many natural coral canyons found throughout the Caribbean and the wider eastern Atlantic Ocean.

Cobia

Family: Rachycentridae

5 COBIA

Rachycentron canadum

Vaguely shark-shaped, the Cobia has quite a long torpedo-like body that grows to around 1.8m (5ft 11in). It has large pectoral and dorsal fins which are extended at all times. It enjoys hanging out under piers and anchored ships, and is easily approached, appearing unafraid of divers.

Snook

Family: Centropomidae

6 COMMON SNOOK

Centropomus undecimalis

The Snook is a large fish resembling cod, and can grow to over 1m (3ft 3in). It has a strong curved lateral line along a greenish-grey body. Usually seen in small groups, it tends to prefer sheltered areas such as mangrove forests, but will venture into the coastal seagrass beds. It is fairly common around the Florida Keys and is known extensively from the north coast of South America and the ABCs.

1

2

3

4

5

6

Chub

Family: Kyphosidae

1 BERMUDA CHUB

Kyphosus sectatrix

This is a large round silvery fish that grows to over 60cm (2ft). It often displays golden stripes or dark blotches over the body, and has dark fins and tail. As the name suggests, it is fairly common from Bermuda south to Brazil and is found throughout the zone.

Spadefish

Family: Ephippidae

2 SPADEFISH

Chaetodipterus faber

Disc-shaped and similar in shape and size to Angelfish, this fish can grow to around 90cm (3ft). It is silvery with same-sized dorsal and anal fins set to the back of the body. There are four or five dark vertical bars down the body and at the base of the tail. It hangs out in the water column above bank reefs in small hunting groups.

Bonefish

Family: Albulidae

3 BONEFISH

Albula vulpes

Popular with sea anglers, this inshore fish grows to around 1m (3ft 3in) long. It has a short underslung mouth and a single, triangular dorsal fin placed midway along the back, and a deeply forked tail. The body is greenish silver. It feeds in small groups on an incoming tide, over seagrass beds and near mangrove forests at depths less than 9m (30ft).

Remoras

Family: Echeneidae

4 WHITEFIN REMORA

Echeneis neucratoides

Very similar to the Sharksucker, this species grows to 75cm (2ft 6in). It is predominantly white with black anal and dorsal fins and tail. It also has a wide black stripe that runs the full length of the body. It attaches itself to large rays, sharks and turtles and can be found in large numbers swimming ahead of Whale Sharks.

5 SHARKSUCKER

Echeneis naucrates

This species has specially adapted dorsal fins which allow it to attach to a host; it is therefore encountered attached to large fish such as sharks and rays; also turtles and even divers (if you get too close to one that is looking for a new home!). Dark grey and black overall, it has a thin white border to the tail and dorsal fin and larger adults show a thin white stripe running from the pointed nose to the end of the anal fin.

Mojarras

Family: Gerreidae

MOJARRAS HAVE LARGE and reflective silver scales, pointed mouths and forked tails. They are rarely found over the coral reefs, instead preferring sandy or coral rubble open areas where they swim in small groups, feeding on invertebrates.

6 YELLOWFIN MOJARRA

Gerres cinereus

Growing to a maximum of around 50cm (1ft 8in), this fish has a row of slightly darker vertical bands down the body. The pectoral fins are tinged with yellow and the end of the tail is much darker. It is usually seen in small groups, feeding on invertebrates on the seabed. It is found throughout the Caribbean zone and Gulf of Mexico.

1

2

3

4

5

6

1 SLENDER MOJARRA

Eucinostomus jonesi

Growing to only around 20cm (8in), this slender fish has silver scales and a large black eye. It has a long dorsal fin with a tall leading edge. It is found all over the entire zone and much of the western Atlantic Ocean.

2 IRISH PAMPANO

Diapterus auratus

More rounded in shape than the previous two species, this fish grows to around 30cm (1ft). It has a high ridged back. The tall front edge of the dorsal fin is pointed and tinged black, as is the anal fin. The silvery body and tail may be tinged with green. It is found throughout the region.

Porgy or Bream

Family: Sparidae

A NUMBER OF THESE SPECIES are very similar to their counterparts in the eastern Atlantic and Mediterranean. They are silvery with a high back, a long dorsal fin and forked tail. The mouth has fleshy lips and is set well below the eyes. They tend to be solitary and are found in all areas.

3 SILVER PORGY

Diplodus argenteus

Growing to around 30cm (1ft), this is a silver, streamlined fish with a black spot on the upper portion of the tail's base that may form a band over both sides. It has a curved lateral line and a shallow forked tail. Although regarded as uncommon, it is found over a wide range and enjoys shallow rougher water.

4 SPOTTAIL PORGY

Diplodus holbrookii

Very similar to the Silver Porgy, this fish grows to 45cm (1ft 6in) and also has a silver body, but with a larger black spot at the base of the tail, which may form a wide band all around. It may be more cream in colour, with cream to ochre stripes along the body and blueish pelvic fins.

5 SAUCEREYE PORGY

Calamus calamus

This species grows to around 40cm (1ft 4in). It has a line below each eye and a high nape. The silver scales are quite reflective and it usually has a yellowish blush to the face and nape. I first encountered this species in Bermuda, where it is quite common.

6 SHEEPSHEAD PORGY

Calamus penna

The Sheepshead Porgy grows to 90cm (3ft). It is easily recognised by the five or six vertical or slightly diagonal dark bars that run the length of the body. There is also a bar at the base of the tail, and the face is a silvery grey. It is very rare in the Caribbean but more common in Florida and the Gulf of Mexico, usually occurring in shallow water less than 12m (39ft).

Goatfish

Family: Mullidae

7 SPOTTED GOATFISH

Pseudupenneus maculatus

The Spotted Goatfish grows to around 30cm (1ft). It comes in a variety of colour forms and can change colour to suit its environment, particularly at night when it becomes many shades of mottled red and pink. It has a row of three dark blotches along its flanks, and long barbels under the chin which it uses to agitate the sandy seabed in the search for food. It is fairly common through the entire region.

1

2

3

4

5

6

7

1 **YELLOW GOATFISH**

Mulloidichthys martinicus

This species is slightly larger and more rounded than the Spotted Goatfish, reaching lengths of over 35cm (1ft 2in). It has a whitish-silver body, with an obvious yellow midline which runs from behind the gill cover to the base of the tail, which is also yellow. It often forms large schools over the reefs and sand flats, and is common in all areas.

Mullet

Family: Muglidae

2 **STRIPED MULLET**

Mugil caphalus

This mullet can reach 60cm (2ft) long and can often be found in huge schools feeding on the surface. Often found in shallow lagoons, this fish is quite wary and will move away from you if approached. It is quite common in the region, except the far east, and enjoys fresh water and brackish conditions as well as the sea.

Pufferfish

Family: Tetradontidae, Diodontidae

PUFFERFISH AND BURRFISH are a curious bunch of fish with the ability to suck in great amounts of water to inflate the body, making them unpalatable and very difficult for predators to eat. They are all found in all areas of the reef, sandy planes and lagoons.

3 **SHARPNOSE PUFFER**

Canthigaster rostrata

This is the most common of all the little pufferfish that is found on most of the coral reefs. It can grow to around 10cm (4in), but is usually much smaller than this. It has blueish stripes which radiate out from the eye, a dark brown upper body and a golden lower half covered in small blue spots. There are dark lines to the top and bottom of the base of the tail, which extend out and over the top and bottom of the tail.

4 **BANDTAIL PUFFER**

Sphoeroides spengleri

This elongated pufferfish grows to around 30cm (1ft). The name comes from the two obvious vertical bands on the rounded tail. It has a brown upper body and white underside with a very obvious line of brown dots along the entire length. Widespread from Bermuda to Brazil, it is fairly common throughout the entire region.

5 **CHECKERED PUFFER**

Sphoeroides testudineus

This puffer is a similar shape and size to the Bandtail Puffer. It has a fawn to grey body with numerous polygonal and randomly shaped blotches over all of the upper body. It often rests in a depression on the seabed.

6 **BALLOONFISH or SPINY PUFFER**

Diodon holocanthus

Often found in pairs on the reef, this fish has a variable tan and darkly blotched body. It has very long spines on the body, which normally face towards the tail but stick out at 90 degrees if the fish feels threatened. Its large eyes assist it when feeding at night.

7 **PORCUPINEFISH**

Diodon histrix

A large, pale pufferfish, growing to over 90cm (3ft), this species has tiny spines all over the body, which stick out when the body inflates. The fins and tail are covered in dark spots.

1

2

3

4

5

6

7

1 WEB BURRFISH

Chilomycterus antillarum

Growing to about 30cm (1ft), this puffer has reticulated markings over all of the body. It has large eyes and mostly golden yellow, rather erect spines which are always on display. It is found in all areas of the reef and coral rubble seabed, and occurs in most regions, but is regarded as rare as it tends to hide away during the day.

Cowfish and Trunkfish

Family: Ostraciidae

2 HONEYCOMB COWFISH

Acanthostracion polygonia

This species grows to around 50cm (1ft 8in). The entire bluish-green body is covered in a honeycomb pattern, with large darker patches on the flanks. Roughly triangular in shape, it has a couple of spiny 'horns' pointed forwards in front of the eyes. These distinguish it from the trunkfish, which have no horns. It is found in all areas of the reef, but rarely deeper than 25m (80ft).

3 SMOOTH TRUNKFISH

Lactophrys triqueter

One of my favourite fish, this trunkfish is triangular with a dark body, covered in light spots that may morph into hexagonal shapes behind the pectoral fins. The fins are dark, as are the lips around a small rounded mouth. Juveniles are tiny round black balls with a small pointed mouth, small eyes and large white spots.

4 SPOTTED TRUNKFISH

Lactophrys bicaudalis

Longer and more rounded than the Smooth Trunkfish, this species grows to around 30cm (1ft). Its white body is covered in tiny dark spots, as are the dorsal and anal fins and tail. The pectoral fins are clear and the snout is white. It may also have three or four white spots on the upper body. It is rather shy, preferring to live in coral recesses, but has a wide range.

Filefish

Family: Monacanthidae

5 SCRAWLED FILEFISH

Aluterus scriptus

Quite large at around 90cm (3ft), this is an elongated green and yellow multi-coloured fish, with random lines of dark spots. It has a long pointed snout and small mouth. The back of the body tapers to a narrow tail base, but the tail itself is large. It has a sharp pointed dorsal spike located just above the eyes.

6 WHITE-SPOTTED FILEFISH

Cantherhines macrocerus

This flattened filefish has a steep forehead and small white mouth. It may be grey on top, ochre below and covered in faint white spots, but these may disappear. Some have a lower golden yellow body and a black tail.

7 ORANGE SPOTTED FILEFISH

Cantherhines pullus

A smaller filefish growing to around 20cm (8in), this species has a greyish-brown body with a series of rows of orange spots which may become lines towards the tail. The tail is yellow and it has a white spot at the top of the tail base.

8 SLENDER FILEFISH

Monocanthus tuckeri

The smallest of the filefish, this species only grows to 10cm (4in) but is usually just half this size. It has an elongated snout and variable colours, though the upperside is darker and the underside white. It also has reticulate markings. The pelvic fins are large and rounded, and there is a dorsal spike above the eye.

1

2

3

4

5

6

7

8

Surgeonfish

Family: Acanthuridae

1 BLUE TANG

Acanthurus coeruleus

This is a common reef fish that grows to around 36cm (1ft 2in). Adults are a lovely blue over the entire body with blue striped fins and tail, and an unmistakable blade or spine at the base of the tail. At night, they may show four paler blue bars. In contrast, the juvenile is brilliant yellow, with a blue-rimmed eye and blue edge to the dorsal and anal fins. As it transitions, the juvenile develops ochre bars down the body. It occurs all over the region, and the western Atlantic.

Juv. 1

2 OCEAN SURGEONFISH

Acanthurus bahianus

Similar-sized to the Blue Tang, this species is more elongated and coloured pale blue to grey and brown. It also has a spiny blade at the base of the tail. Often found in loose aggregations and often mixing with Blue Tangs, it is found throughout the entire region.

Triggerfish

Family: Balistidae

TRIGGERFISH THE WORLD OVER lay their eggs in a depression amongst coral rubble or rough sand, close to healthy reefs. They vigorously defend their territory and will attack anything or anyone that gets too close.

3 QUEEN TRIGGERFISH

Balistes vetula

This is an incredibly colourful fish in all stages of its development. It grows to around 60cm (2ft). Its colours are paler towards the face and more blue-purple towards the tail. It has radiating lines that spread out from the eye, and two large neon blue-silver lines that run around the snout to the gill covers; it also has rows of small spots which form lines down the flanks. The dorsal and anal fins are equal in size.

4 BLACK DURGON

Melichthys niger

An under-appreciated and beautifully coloured fish, this species grows to around 30cm (1ft). It is mainly blue, with an ochre blush to the cheeks. It also has blue radiating lines which spread from the eye forwards and over the nape. There is a pale blue line at the base of the dorsal and anal fins, and each scale on the body appears edged in black. It is circumtropical and is common throughout the entire region, often occurring in large numbers over the reef.

5 OCEAN TRIGGERFISH

Canthidermis sufflamen

This relatively drab triggerfish can grow to 60cm (2ft). It is dull grey, sometimes blotched with darker or lighter shades of grey, and can change colour quickly. It has a distinctive dark spot at the base of the pectoral fin and both the anal and dorsal fins are placed well back on the body. It is often found in small groups in midwater over offshore coral banks.

1
2
3
4
5

Porkfish

Family: Haemulidae

1 PORKFISH

Anisotremus virginicus

Growing to around 30cm (1ft), the Porkfish goes through a number of colour changes from juvenile to adult. The youngest have a yellow face with the front dorsal fin and pelvic fins yellow. The tail is clear, and there is a black spot on the base of the tail and two black horizontal stripes along the white body. With age they lose the black spot and stripes, gain a yellow stripe and two black diagonal bars on the head, one of which is through the eye. When fully adult, they are more yellow overall with yellow-silver stripes.

Snappers

Families: Haemulidae, Lutjanidae

THIS MASSIVE CIRCUMTROPICAL family is named for the way they snap their jaws when hooked by a fisherman! Most are around 60cm (2ft) long, with large fleshy lips, long dorsal fins and usually a steep forehead.

2 DOGTOOTH SNAPPER

Lutjanus jocu

The two long fang-like teeth are very obvious and give this fish its common name. It grows to 90cm (3ft) and can be grey to reddish brown. It may be found alone or in large groups mixed with other snappers. It is fairly common everywhere on all types of reefs.

3 MAHOGANY SNAPPER

Lutjanus mahogoni

Growing to around 30cm (1ft) or slightly over, this snapper is mahogany when adult, but juveniles and intermediate stages are white with a mahogany line on the end of the tail and dorsal fin. The pelvic and anal fins are yellow and there is a distinctive black spot towards the rear of the white body. It is fairly common throughout the entire region.

4 LANE SNAPPER

Lutjanus synagris

The Lane Snapper is similar to the Mahogany Snapper, though larger, growing to 45cm (1ft 3in). The elongated body is silvery white with yellow on the pelvic and anal fins. It has a black spot to the rear of the body. It is fairly common everywhere and on all types of reefs, usually in large numbers.

5 YELLOWTAIL SNAPPER

Ocyurus chrysurus

This elongated silvery fish grows to over 60cm (2ft). It has a yellow dorsal fin, and a wide yellow stripe down the centre of the body which terminates at a yellow tail. Common mid-water above all of the Caribbean reefs, it tends to be in fairly large schools.

6 SCHOOLMASTER

Lutjanus apodus

One of the most common snappers found on the Caribbean reefs, this species grows to 60cm (2ft) and has a silvery body, a fairly pointed snout with large eyes, and all of the fins and tail are yellow. It is especially common on spur and groove reefs, often occurring in fairly large numbers and hiding out under coral overhangs.

1

Juv. 3

2

4

5

6

Grunts

Family: Haemulidae

GRUNTS ARE NAMED for the noise they make when they grind their teeth, and the noise is apparently amplified by their swim bladders. Very similar to snappers in shape and size, they tend to be more colourful. They hunt at night over the seagrass beds and amongst mangrove roots.

1 FRENCH GRUNT

Haemulon flavolineatum

This species has a series of small diagonal stripes just behind the pectoral fins and above the lateral line. The body is blue-silver with yellow stripes, or arguably yellow with blue-silver stripes! All the fins and tail are yellow. This fish can be found in large numbers on most reefs throughout the region.

2 BLUESTRIPED GRUNT

Haemulon sciurus

Very similar to the French Grunt, this fish grows to 45cm (1ft 6in). The rear of the dorsal fin and tail are black with a lighter outer edge, and the snout has diagonal silvery blue lines that run from the large eye to the mouth with its large fleshy lips.

3 CAESAR GRUNT

Haemulon carbonarium

This smaller species grows to just over 30cm (1ft). It is silvery yellow with a dark anal fin, tail and rear of the dorsal fin. It has obvious coppery bronze stripes along the length of the body. It can be found in small schools, often mixed in with Cottonwicks.

4 COTTONWICK

Haemulon melanurum

The Cottonwick is quite easy to identify, being white with a black lateral line that runs from the snout through the eye to just beyond the gill covers, and extends diagonally to the tail. The dorsal fin is broad and black, and it and the tail are edged white. This common fish is often found in large schools on a fairly shallow reef less than 15m (50ft), particularly around the offshore reefs above Cozumel.

5 WHITE MARGATE

Haemulon album

A very common fish over the entire range, this species grows to over 60cm (2ft). It is silver-white with obvious scales. The back of the dorsal fin and end of the tail are grey. It has a high-arched back and a small white eye. Large schools are often found over mid to shallow reefs and seagrass beds.

6 LATIN GRUNT

Haemulon steindachneri

A smaller grunt, barely reaching 30cm (1ft), this fish is marked with oblique diagonal stripes and yellowish longitudinal stripes. It has a pointed snout and a large black or mahogany blotch near the base of the dusky tail. Rare in the range, but more common in the south, it is often in small schools amongst other grunts and snappers.

7 SMALLMOUTH GRUNT

Haemulon chrysargyreum

This grunt grows to around 26cm (10in). It has five wide yellow stripes over a silvery-blue body and all of the fins are yellow. It can be found on most types of coral reef, and forms large schools in southern Florida.

1

2

3

4

5

6

7

1 SPANISH GRUNT

Haemulon macrostomum

Darker than other grunts, this species grows to around 46cm (1ft 6in). It has wide dark stripes along the upper half of the body and a yellowish tinge to the fins and tail in its adult phase. The juvenile also has wide dark stripes and a lower line midway down the body that passes from the snout, through the eye to the base of the tail, where there is a wedge-shaped black mark. It is widespread, and often mixes with other grunts.

2 WHITE GRUNT

Haemulon plumierii

This larger, iridescent pale grunt grows up to 53cm (1ft 9in). It has radiating fine stripes from the snout, past the eye. It has a strong yellowish lateral line, and all of the scales are highlighted with either blue or yellow. It is found throughout the entire region, always amongst coral reefs in less than 12m (39ft).

3 TOMTATE

Haemulon aurolineatum

An elongated fish, the Tomtate only grows to around 20cm (8in). It has bold golden stripes which run from the snout to the base of the forked tail, where there is a large black blotch. It is often seen in fairly large schools close to the bottom in coral valleys throughout the range.

4 BLACK MARGATE

Anisotremus surinamensis

A dark fish with a high back, this species also has a large dark smudge behind the gill covers that covers half the body, and black fins and tail. A shy fish, it hangs around the entrance of caves and caverns, usually in small groups.

Parrotfish

Family: Scaridae

PARROTFISH ARE DISTANT cousins of wrasse but, rather than fleshy lips, parrotfish have a strong, calcified beak and wide molar plate. This enables them to bite off chunks of coral and grind them up to get at the soft polyps within, in the process contributing significantly to production of sand around the offshore reefs. Parrotfish are voracious eaters and are always found in small groups of perhaps ten to twenty females with one large 'super-male'.

5 BLUE PARROTFISH

Scarus coeruleus

The Blue Parrotfish grows to around 120cm (4ft). It has a sharp, squared off head, a prominent mouth and long nape. Adults are blue, while juveniles have a yellow blush to the nape. Rarely seen on the reefs, it is often found over sand and at night will rest amidst the corals, sea fans and sponges.

6 MIDNIGHT PARROTFISH

Scarus coelestinus

Growing to around 90cm (3ft), this parrotfish can often be found in fairly large schools. All of the colour phases are blue, but with pale blue markings around the head. The scales are highlighted with a mid-blue colour. I have found them most often in Florida, and south to the Bahamas.

1 RAINBOW PARROTFISH

Scarus guacamaia

The largest parrotfish in this region, this species can grow to around 1.8m (5ft 11in). It has an orange-brown head, blending into an emerald green body with a bronze tail. As they get older, the adults may acquire an alga moustache! I have never seen it on open coral reefs during daylight hours, but have found it several times at night in the Bahamas, where it comes to rest on some of the wrecks. It is usually seen in pairs and occurs from Bermuda to Brazil.

2 QUEEN PARROTFISH

Scarus vetula

This common parrotfish grows to 60cm (2ft). It has very obvious large blue stripes around the mouth and the pectoral fin has a light blue stripe; it may have a greenish head. Juveniles are more brown-grey with a wide pale grey stripe that leads to the brown tail. It can be found on most coral reefs and at night can be found resting in a coral alcove, where it produces a mucus cocoon for protection against predatory eels (below). It is found over the entire range.

3 REDTAIL PARROTFISH

Sparisoma chrysopterum

The Redtail Parrotfish grows to 49cm (1ft 7in). It is more green and blue than other species and has a red crescent on the tail's outer edge, as well as on the dorsal fin. It ranges from Bermuda to Brazil but its distribution is patchy.

4 PRINCESS PARROTFISH

Scarus taeniopterus

Growing to just over 30cm (1ft), the adult of this species is easily recognised by the wide and smudged yellow stripe that runs from the pectoral fin to near the base of the tail. Juveniles, however, are similar to juveniles of other species. It is fairly common and found in all regions.

5 STRIPED PARROTFISH

Scarus iseri

Almost identical in appearance and size to the Princess Parrotfish, this species is distinguished by a yellow snout and clear fins and tail. It is common throughout the range.

6 STOPLIGHT PARROTFISH

Sparisoma viride

A larger parrotfish, growing to 64cm (2ft 1in), this species is fairly common on most reefs and often confused with the Queen Parrotfish, which has similar coloration on the head and snout. There is an obvious yellow spot at the top of the gill covers, which often have a yellow to pink outer lining. The tail also has yellow markings and the broad stripe running back from the mouth makes it look like it is grinning! The juveniles are almost entirely red.

7 REDBAND PARROTFISH

Sparisoma aurofrenatum

This is another parrotfish which undertakes several colour changes, making identification often tiresome! Juveniles are metallic greenish with light purplish fins and tail. They then mature into an 'initial phase', which is pictured on the right. Adults reach around 30cm (1ft) and resemble some other species, with few identifying features other than a purple hue to the edges of the tail and fins and a few small black and yellow spots behind the eye. They also have a diagonal stripe that runs above the gill covers from the mouth.

1

2

3

4

5

6

7

Hogfish, Wrasse and Razorfish

Family Labridae

WRASSE FORM A LARGE and diverse family and occur throughout all of the world's oceans. Most species are carnivorous, living on small crustaceans and molluscs, and a few have very strong jaws with which to crush sea urchins. They are found around all coastlines, quite often in very large numbers. All are around three times longer than they are tall, and have large, fleshy, protruding lips, and prominent eyes set high on the head. They have one single low dorsal fin and short, rounded pectoral fins. Juveniles of some species act as cleaners to other fish. Most go through various colour changes as they reach maturity and are often difficult to identify. Razorfish are known for their sharp snout and steep forehead, enabling them to dive head first into patches of soft sand when danger threatens.

1 HOGFISH

Lachnolaimus maximus

Growing to around 90cm (3ft), the Hogfish has three long spines at the front of the dorsal fin. The fin then tapers away before becoming longer again at the tail, which has a dark band. It has a steeply sloped forehead and a wide brown stripe that runs from the upper lip to the dorsal fin. Adults are usually solitary and roam the reef and sandy or coral rubble seabed, rooting about for invertebrates to eat.

2 SPANISH HOGFISH

Bodianus rufus

A common fish, this species grows to 60cm (2ft). It is distinctive, with a golden-yellow lower body and tail and light purple upperside. It is known to act as a cleaner, actively removing parasites from various species of fish including jacks and groupers.

3 SPOTFIN HOGFISH

Bodianus pulchellus

Only growing to around 20cm (8in), this wrasse is easily identified with its bright red to purple fore body and golden-yellow rear, and its red eye. It acts as a cleaner species for other much larger fish, and can be locally common in some areas, such as the Gulf of Mexico, though rarer in other areas of the Caribbean.

4 CREOLE WRASSE

Clepticus parrae

The Creole Wrasse grows to a maximum of 30cm (1ft). It can be multicoloured with white lips and a blue-black or purple body, leading to a yellowish tail with pink highlights. It prefers deeper reefs or submarine walls, and is often found in large schools along the edge of the reef.

5 PUDDINGWIFE

Halichoeres radiatus

The Puddingwife can grow to 45cm (1ft 6in). It is greenish with some paler highlights, blotches or bars, and may have a yellowish margin to its tail. Juveniles have a dark upper body towards the rear of the dorsal fin, with two white bars. The rest of the body has gold and green markings which form irregular stripes along the flanks. It is fairly common along the western Atlantic seaboard and throughout the Caribbean.

Juv. 5

1

2

3

4

5

1 YELLOWHEAD WRASSE

Halichoeres garnoti

This wrasse grows to around 20cm (8in). Only adults have a yellow head – in earlier stages the body is mainly yellow with a bluish upperside, but at all ages there are a couple of black radiating lines that extend back from the eye. These lines are short but may continue as dots towards the rounded tail. It is found over the entire range.

2 SLIPPERY DICK

Halichoeres bivittatus

Growing to around 20cm (8in), this pale but multicoloured wrasse may have short pale bars down the body, a darker mid lateral line, and a dark spot just above the pectoral fin. Its barred markings may be in various pale colours, but all conform to the same intricate pattern. It is found in sheltered reef and seagrass beds, often with groups with all of the colour phases represented, and occurs across the whole region.

3 BLUEHEAD

Thalassoma bifasciatum

Probably the most common wrasse in the entire region, this species grows to 15cm (6in), and goes through many colour changes as it matures. The first juvenile phase is yellow with a lighter underside and a black spot on the front of the dorsal fin, or white with a wide black midline and top of the body. Some individuals eventually mature to become a 'super-male' with a harem – this terminal phase has a blue head with a black bar behind, followed by a white 'V' shaped bar, then another black bar and the rest of the body a paler greenish blue. The juveniles are commonly seen in small schools and are very active.

4 GREEN RAZORFISH

Xyrichthys splendens

A small fish growing to just over 10cm (4in), the Green Razorfish adult has a red-rimmed eye. The body can be a variety of greenish colours with bars, with a very distinctive black spot midway down the flank. Juveniles resemble bits of weed and have separate filaments at the front of the dorsal fin. The species is found through the entire region, invariably on shallow sandy seabeds and lagoons. It will curl its body around seagrasses and algae.

5 PEARLY RAZORFISH

Xyrichthys novacula

This species grows to 15cm (6in). It has a black eye and there may be various bars down the length of the body. Most are whitish with a little red coloration, but occasional individuals are predominantly red. It is common in the zone and lives on sandy seabeds and seagrass areas.

1

2

Juv. 3

4

5

OTHER VERTEBRATES

Sea Turtles

Superfamily: Chelonioidea

SEA TURTLES ARE regularly seen within the entire Caribbean and Gulf of Mexico region, some more often than others. Long hunted by the indigenous Caribbean peoples, they are now protected under CITES and sea turtle products may not be traded internationally. However, they are still hunted in some areas and Green Turtles are even raised commercially for their meat. Turtles are predated by sharks and are susceptible to boat strikes whilst they are on the surface. They are also often caught as by-catch by commercial fisheries, but the main concern for their future centres on habitat loss and plastic pollution.

1 LOGGERHEAD TURTLE

Caretta caretta

This sea turtle can weigh as much as 181kg (400lb) and typically the carapace is around 1m (3ft 3in) long. It is slightly heart shaped with no overlapping scutes. One of the main identifying pointers is its very large head, accommodating very strong jaws for eating hard-shelled invertebrates. The largest nesting area is in Florida and it is also known to nest in the Cayman Islands, Bahamas, Cuba and Mexico.

2 OLIVE RIDLEY TURTLE

Lepidochelys olivacea

This is one of the world's smallest sea turtles, reaching around 35–50kg (75–110lb) and 71cm (2ft 4in) long when adult. It has on average nine scutes which do not overlap, and gets its name from the olive-green colour to its shell. It reaches sexual maturity at around 12 years old, and is known to nest in Trinidad and Tobago, Curaçao and Panama.

3 GREEN TURTLE

Chelonia mydas

One of the larger sea turtles found in the Caribbean region, its carapace is usually around 1m (3ft 3in) long and it weighs in at 181kg (400lb). Recorded from more than 80 countries, it is found globally in all tropical waters. The scutes are rounded and non-overlapping. There is a single claw on each flipper and the head is rounded with two prefrontal scales between the eyes.
It is known to nest in Costa Rica, Mexico, Florida, Venezuela and Cuba.

4 HAWKSBILL TURTLE

Eretmochelys imbricata

A small sea turtle weighing around 27–90kg (60–200lb) and with a carapace of up to 1m (3ft 3in) long, this is the most commonly encountered species throughout the entire region and is most often found around coral reefs. Hawksbills are known as 'spongivores', having a diet principally of sponges, but they will also feed on other invertebrates. They have two pairs of prefrontal scales between the eyes, and overlapping, pointed scutes on the carapace. They are known to nest on Mona Island (Puerto Rico), Barbados, Mexico, Panama and the US Virgin Islands.

1

2

3

4

Manatee

Family: Trichechidae

1 WEST INDIAN MANATEE

Trichechus manatus

This manatee, also known as a 'sea cow', can measure up to 4m (13ft) long, and weigh as much as 590kg (1,300lb). Unlike cetaceans or sharks, it has paddle-like tail fins, and is a slow-moving herbivore. Its nearest terrestrial relatives are the elephants. Although fully aquatic, it still has vestigial toenails on its forelimbs or flippers. The best place to have friendly encounters with manatees is in the Crystal River area of north-west Florida, where hundreds overwinter and enjoy the constant warm water from the underground springs.

1

Whales

Families: Physeteridae, Balaenopteridae

CETACEANS HAVE A SHARED ancestry with hoofed land mammals, but are adapted to an entirely marine lifestyle. The group includes all whales, dolphins and porpoises. All have two front flippers and no hind limbs, and their tails are unique horizontal extensions to the body, called flukes.

2 HUMPBACK WHALE

Megaptera novaeangliae

Travelling into the Caribbean after bypassing Bermuda, large numbers of Humpback Whales congregate off the Silver Banks, north of the Dominican Republic. The warm and shallow Silver Banks are perfectly suited for them to mate and give birth.

3 MINKE WHALE

Balaenoptera acutorostrata

The Minke Whale is surprisingly common in the Caribbean region and tends to favour coastal locations, whether on the mainland or offshore islands, where upwellings of nutrients attract larger schools of fish.

4 SPERM WHALE

Physeter macrocephalus

These huge-headed whales can stay underwater for more than 20 minutes whilst searching for food, and their massive 4m (13ft) long flukes can propel them into the depths quickly and efficiently. Dominica is now world-renowned for positive encounters with Sperm Whales during the winter months.

2

3

4

Dolphins

Family: Delphinidae

DOLPHINS ARE ACTIVE and often very fast-moving small cetaceans. They are usually highly social.

1 BOTTLENOSE DOLPHIN

Tursiops truncatus

This is the most commonly encountered dolphin in the region. It has a friendly expression, a long beak and high forehead, and grows up to 4m (13ft) long. It enjoys riding the bow wave of fast boats. Pods of Bottlenose Dolphins can be found in many areas, from Bermuda to Brazil and all points in between.

2 FRASER'S DOLPHIN

Lagenodelphis hosei

This species was not fully studied and named until 1971. A small dolphin, it grows to around 3m (10ft) and is quite stocky, with a short beak, small fins and tail. It is greyish on top, with a dirty cream underside running from the eye to near the tail. It forms medium to large groups and is very active on the surface. This group was photographed off Dominica.

3 ATLANTIC SPOTTED DOLPHIN

Stenella frontalis

This species has been observed and studied off Bimini in the northern Bahamas for many years. It has a complex social structure and is known to appear to enjoy human interaction and to play with toys. It lives in small family groups, which actively hunt together at night.

4 FALSE KILLER WHALE

Pseudorca crassidens

Growing to around 6m (20ft), this is a globally rare species of large dolphin that is fairly frequently seen around the eastern Caribbean. Strandings have occurred in some areas. These animals were photographed off Dominica.

5 GREY'S SPINNER DOLPHIN

Stenella longirostris longirostris

Adults are relatively small at around 2m (6ft 6in) and form large family groups, from several dozen individuals to many hundreds, which can cut an amazing swathe when all travelling together on the surface. They are tricoloured, with dark grey on top, cream underneath, and light grey in between. As the name suggests, they are known to leap from the surface and spin many times before crashing back below the waves.

1

2

3

4

5

INVERTEBRATES

CNIDARIANS

Phylum: Cnidaria

THIS IS A LARGE GROUP of fundamentally simple animals, widely distributed and seemingly highly diverse in form. The phylum includes true hard corals, soft corals, sea pens, sea fans, anemones, sea firs (hydroids), zoanthids and jellyfish. They come in two different forms – attached polyps, and free-floating medusae.

All species are radially symmetrical, without a right or left side. They have a single body cavity with a single opening, usually surrounded by one or numerous rings of tentacles. It is these tentacles with their various forms and functions that are common to all members of the phylum. Each tentacle is armed with stinging cells, cnidocytes, containing specialised organelles called nematocysts or cnidocysts. Nematocysts are small capsules which contain a smooth or barbed thread coiled inside. Used for defence or aggression and for catching prey, these stinging cells may also be found on other parts of the body. The toxins released when these 'harpoons' are fired may be powerful enough to cause a severe sting.

Anemones

Class: Hexacorallia

THIS SUBDIVISION OF CNIDARIA is the largest group and includes the anemones and related species. The name Hexacorallia indicates that the polyp tentacles are arranged in groups of six, rather than eight as in some other cnidarians. Most anemones are solitary and attached to a hard substrate by means of a basal sucker. Others burrow into soft sand or mud, creating a protective tube into which they can withdraw. Some species live in close proximity because of their budding habit, while others produce larvae. Some reproduce through basal laceration, where part of the column or base is detached and grows into a new polyp. All anemones are carnivorous, trapping unwary small fish or invertebrates within the sticky grasp of their tentacles.

1 GIANT ANEMONE

Condylactis gigantea

The largest anemone in the entire region, this species is common everywhere. Similar in colour to the Club-tipped Anemone, it is much larger overall, growing to over 30cm (1ft). The tentacles are usually pale, with large, mauve club-shaped tips. This anemone has a symbiotic relationship with several species, including the Diamond Blenny (see page 58), Pederson's Cleaner Shrimp (see page 156), Squat Anemone Shrimp (see page 154) and Banded Clinging Crab (see page 146).

2 SUN ANEMONE

Stichodactyla helianthus

Distinguished by its hundreds of short, thick tentacles, this anemone is generally a shade of green. It can grow to 15cm (6in) and is found in much of the central and eastern Caribbean. It too has a number of diverse symbionts.

1

2

1 ELEGANT ANEMONE

Actinoporus elegans

A plate-shaped anemone, this species grows to 23cm (9in) and has many tiny, knobby tentacles. It is usually pale grey and mottled cream, the colours often arranged in bands. More common in the central and western Caribbean, it prefers to live in rough coral rubble areas and rough sand, where it can retract into the seabed if there is any danger.

2 CORKSCREW ANEMONE

Bartholomea annulata

Very similar to the Knobby Anemone, this species has many thin and pointed translucent tentacles which have white corkscrew-shaped tips. It is otherwise grey and white and occurs amongst coral rubble or at the edge of a reef. Common throughout the entire range, it is usually accompanied by its commensal partners: the Scarlet Pistol Shrimp (see page 158) and Pederson's Cleaner Shrimp (see page 156).

3 KNOBBY ANEMONE

Laviactis lucida

Similar to the Corkscrew Anemone, this species only grows to around 10cm (4in) across. It has translucent tentacles with obvious nematocysts that appear as little knobs. It is common everywhere.

4 BANDED TUBE-DWELLING ANEMONE

Arachnanthus nocturnus

This anemone's ring of tentacles is banded brown and cream, and the main tube is usually pale cream or white. As the scientific name would suggest, it is nocturnal, hiding in soft sand or mud during the day and then extending out to catch plankton at night.

5 GIANT TUBE-DWELLING ANEMONE

Ceriantheomorphe brasiliensis

With a span of around 30cm (1ft), this somewhat rare anemone has several rings of tentacles, which are usually greyish or lavender-coloured, and are fully retractile. It comes out at night and, whilst it is found throughout the range, it is seldom observed.

6 'LAVENDER TUBE-DWELLING ANEMONE'

Unidentified sp.

This may just be a variant of another tube-dwelling anemone species, but the opaque and unusually coloured tentacles appear to set it apart from the other species. Also nocturnal, it is found in the central to southern Caribbean, as far as Venezuela.

7 BERRIED ANEMONE or STINGING ALICIA

Alicia mirabilis

This species is sometimes referred to as the 'wonder anemone', owing to its amazing stretching ability. During the day, it closes up and becomes a small inconspicuous lump, standing only 10–15cm (4–6in) tall with its tentacles retracted. At night it can stretch to over 1m (3ft 3in). It is identified by the numerous small berries or tubercles that cover its body; these are armed with stinging cells and can be quite harmful to humans if touched. By night, its long tentacles trap prey in the current. It occurs from Bermuda to Brazil, through the Atlantic and Caribbean and even the Mediterranean.

1

2

3

4

5

6

7

1 BRANCHING ANEMONE

Lebrunia danae

This is an olive-green anemone with pale markings. Each flattened tentacle branches into two and then may split further. It has a couple of friends in Pederson's Cleaner Shrimps and the Sun Anemone Shrimp. Whilst it is regarded as common in the range, I have only found it in the Windward Islands.

2 HITCHHIKING ANEMONE

Calliactis tricolor

Like all oceans and seas on the planet, there are hermit crabs which choose to carry anemones around as protection. The anemones benefit from a supply of food scraps, as hermit crabs are invariably messy eaters! One crab may carry three or four anemones at once. The anemones emit sticky threads loaded with nematocysts if danger threatens.

3 CLUB-TIPPED ANEMONE

Telmatctis cricoides

This species can grow to 10cm (4in) or larger. As the name suggests, it has obvious clubs that tip the tentacles. The disc is flattened and the main body is generally hidden in a rocky crevice. It is fairly common in much of the Atlantic and around the Azores and Madeira, and whilst not reported in Florida, it occurs in central and southern Caribbean areas.

4 TURTLE GRASS ANEMONE

Bunodeopsis globulifera

The feeding tentacles of this small green anemone only come out at night. During the day, the anemone forms a small flattened ovoid with small tubercles. Growing on Turtle Grass (see page 208) leaves, it can be locally common in a number of areas. The tentacles are packed with stinging nematocysts which can produce painful blisters that may require medical attention, so do not touch.

5 BROWN ANEMONE

Exaiptasia diaphana

Another small anemone, this species only grows to around 2cm (¾in). It is found around seagrass beds and only extends its tentacles to feed at night. It prefers shallow, protected lagoons and waters shallower than 6m (20ft).

Corallimorphians

Order: Corellimorpha

OFTEN MISTAKEN for anemones, corallimorphians have a distinctive flattened disc and an extended mouth, with their small tentacles extending out like the spokes of a wheel.

6 ORANGE BALL CORALLIMORPH

Pseudocorynactis caibbeorum

This species can extend to 5cm (2in). It may be almost fluorescent with a yellow body and translucent tentacles, each tipped with a bright orange club. Only ever seen at night, it is more usually found where corals or coral rubble adjoin fine sand. This attractive species is a sought-after find on the Caribbean reefs.

7 FORKED TENTACLE CORALLIMORPH

Discosoma carlgreni

The oral disc of this species can grow to over 5cm (2in). It has widely spaced tentacles in a multitude of colours, but mainly green on green, or aquamarine. The small tentacles can fork a number of times. Often, several individuals cluster together, vying for space.

8 WARTY CORALLIMORPH

Discosoma osculifera

Similar to the previous species, this corallimorph is larger, growing to almost 10cm (4in) across the oral disc. The stumpy tentacles can inflate and often resemble fat warts, hence the name. It is fairly widespread throughout the entire zone and north-east as far as Bermuda, where it is quite common.

1
2
3
4
5
6
7
8

Zoanthids

Order: Zoantharia

THESE ARE VERY SMALL cnidarians that have a small oral disc with two sets of tentacles around its margin. Most form large colonies, and tend to inhabit various species of sponge.

1 SPONGE ZOANTHID

Parazoanthus parasiticus

This species will heavily encrust several species of sponge, and its small yellowish tentacles on a pale body are very obvious. As the scientific name suggests, it was formerly regarded as a parasitic zoanthid. However, it does not appear to exhibit parasitic behaviour, instead living in commensal harmony with its host. It is common through the entire range.

2 GOLDEN ZOANTHID

Parazoanthus swiftii

One of the most striking of all the zoanthids, this brilliant golden variety is small with polyps measuring less than 1cm (⅓in). It encrusts several species of sponge, but is not necessarily exclusive to any particular type, and will encrust whole lengths of rope sponge.

3 BROWN SPONGE ZOANTHID

Epizoanthus cutressi

This tiny colonial zoanthid grows to less than 1cm (⅓in). It has 20 cream tentacles around the edge of the brown oral disc. The polyps appear to grow in rows over the host sponge.

4 MAT ZOANTHID

Zoanthus pulchellus

A small zoanthid, this species has a 1.5cm (½in) oral disc. There can be many hundreds of polyps together, forming large mats over a smooth surface such as under a pier, or around some of the smoother-surfaced corals. This species is common throughout the entire region and can be very numerous in some places.

5 SOLITARY ZOANTHID

***Palythoa* sp.**

Very similar to the Sun Zoanthid in size and coloration, this undescribed species has very long oral tentacles around the rim of the oral disc and an extended whitish body. It is widespread, but often overlooked due to its small size.

6 SUN ZOANTHID

Palythoa grandis

This zoanthid's oral disc is around 2cm (¾in). The oral disc is very flattened and usually coloured in shades of brown, cream and green. Whilst large numbers can be found together, they are never tightly packed together in a mat. It is relatively common through Florida and into the Gulf of Mexico and Caribbean.

7 BROWN ZOANTHID

***Zoanthus* sp.**

Each polyp only averages around 1cm (⅓in) across, and forms small mat-like colonies or clusters. It is creamish brown with a flat disc and two rings of tentacles, which are uniform in colour but have a light spot at each base. Although it has not yet been scientifically described, it is known to be quite common in the region.

8 HYDROID ZOANTHID

Hydrozoanthus tunicans

This species almost always encrusts the Feather Hydroid (*Gymnangium longicauda*), which is harmful to humans. Resembling small anemones which extend up the stems of the hydroid, the polyps range from golden brown through shades of cream. It is typically common through the entire range.

1
2
3
4
5
6
7
8

Corals

Subphylum: Anthozoa

THE CORALS OF THE western Atlantic, Gulf of Mexico and greater Caribbean are many and varied, from small single polyps to simply massive hard corals, such as the brain corals found in Tobago that are the largest in the world. They are the building blocks of the coral reefs and their many and varied forms provide shelter for more than 25 per cent of all marine life. As a coral colony grows, the polyps produce an underlying skeleton of calcium carbonate, with the living animals just a thin veneer on the surface. Each tiny coral polyp has a central mouth and a ring of feeding tentacles. Often the only way to identify many of the similar species is by close-up or macro photography, so I have presented a few here in this way to help.

1 ELKHORN CORAL

Acropora palmata

One of the largest shallow-water corals, this species thrives in well-aerated clean water on bank reefs and inshore fringing reefs. Taking the full brunt of many hurricanes, its reefs can be smashed to pieces, but the wide branching plates it produces are very fast-growing and can quickly regenerate. It was these corals which were instrumental in wrecking so much of the Spanish treasure fleet, due to their shallow-water habitat, making them highly hazardous to ancient shipping.

2 STAGHORN CORAL

Acropora cervicornis

As the name suggests, this multi-branching coral resembles the antlers of deer. Colonies can reach around 3m (10ft) in height and are usually creamy-brown, with protruding corallites containing the feeding polyps. This coral prefers more sheltered waters due to its rather fragile structure. Huge patches can be found off Klein Bonaire.

3 PILLAR CORAL

Dendrogyra cylindrus hyades

Forming large colonies with many cylindrical spires, always facing up towards the surface, this coral's colonies can reach over 3m (10ft) tall. Some can be just large clumps with a wide base, whilst others, such as those on Curaçao, form enormous colonies that cover a huge area where conditions are perfect. It is more common in southern latitudes.

4 YELLOW FINGER CORAL

Madracis mirabilis

This coral forms small, densely packed clumps of upright coral fingers. They are tinged yellow or cream, with fuzzy white polyps. Aggregations may be around 120cm (4ft) across and are quite fragile. This species is found in all of the region in all depths, but always in clean aerated water.

5 FINGER CORAL

Porites porites

This species forms largeish clumps of many rounded, branching arms, and can grow to over 120cm (4ft) across. It is fairly common throughout the entire range and is found in most depths, but particularly often at the edge of fringing reefs and bank reefs.

6 BOULDER STAR CORAL

Orbicella annularis

There are four different morphs of this large reef-building coral, and although they differ, their shared characteristics include an uneven surface, often with knobby protrusions. The corallites are raised and have paler centres, with the feeding polyps only appearing at night.

7 GREAT STAR CORAL

Montastraea cavernosa

A large coral with smooth corallites, this species comes in many different colours from near red to green, ochre and cream, with different-coloured stripes on the sides. It can grow to around 3m (10ft), and is found in all of the range, often rising high up on the reef crest. It is commonly used by a large number of cleaning gobies as shelter.

1
2
3
4
5
6
7

1 ROUGH STAR CORAL

Isophyllastrea rigida

Coming in a number of colours from green to grey, this species has a large, ridged, fleshy lip to the larger valleys, which hold one or two polyps each. It forms small domes, which may be over 2m (6ft 6in) high. It is usually found in clear, clean, well-aerated waters on the reef crest, and occurs in most regions.

2 ELLIPTICAL or PINEAPPLE STAR CORAL

Dichocoenia stokesi

Fairly common throughout the Caribbean range, this green to cream-coloured coral forms small elongated domes up to 35cm (1ft 2in). The most obvious characteristic is the raised corallites which are elliptical, Y-shaped or circular, all bunched together. It is found lower on the reef.

3 SMOOTH FLOWER CORAL

Eusmilia fastigiata

Forming small clumps around 60cm (2ft), this coral's polyps are on long separate stalks and can be blue-grey to cream. Often the polyps are elongated or elliptical and have very obvious raised ridges. It is found in all areas of the Caribbean and western Atlantic.

4 TUBE CORAL

Cladocora arbuscula

This small species grows to around 15cm (6in). Common through the entire range, especially in shallow seagrass beds, this coral appears to enjoy turbidity and muddy waters. It grows in broken-looking colonies in lagoons and inner reef flats.

5 HIDDEN CUP CORAL

Phyllangia americana americana

Often difficult to spot due to the polyps being only around 1cm (⅓in), this coral is also often overgrown by sponges. It is variably brown and cream, and has sunken central pits with a raised ridge consisting of six raised and six smaller septa around the rim. It prefers low light in caverns and overhangs.

6 ORANGE CUP CORAL

Tubastrea coccinea

This Indo-Pacific species grows to only 2cm (¾in). The body and tentacles are a bright gold and it forms small clumps in shaded areas, with often larger congregations under overhanging reefs. It extends its feeding polyps at night. The species was first described outside its normal range in the Red Sea in 1829, and was later discovered in Acapulco, Mexico, in 1866, seemingly transferred by attachment to ships' hulls. It probably entered into the Caribbean with the opening of the Panama Canal in 1914, and was recorded in Curaçao in 1943. From there it was found in Jamaica, Cuba and the Cayman Islands by 1955. By 1970 the eastward spread of this unassuming little golden polyp had reached the Virgin Islands, and then the Bahamas and Turks and Caicos in 1985. By 1999, it was prevalent in all areas of the Caribbean and Gulf of Mexico.

7 SYMMETRICAL BRAIN CORAL

Diploria strigosa

This coral's domes can reach 2m (6ft 6in). Many are a nice cylindrical shape, but some are quite convoluted and may grow over other dead coral formations. The valleys or grooves can be quite long and may become interconnected. The ridges are smooth and green to cream in colour. They are very common everywhere.

8 KNOBBY BRAIN CORAL

Pseudodiploria clivosa

Common in all areas, this coral often encrusts dead corals or rocky substrates and rarely forms the classic dome shape of other brain corals. All of its ridges are sharply defined and tightly packed, leaving no room for the classic brain coral grooves.

1
2
3
4
5
6
7
8

1 BOULDER BRAIN CORAL

Colpophyllia natans

A common coral which grows as extraordinarily large domes in the Speyside area of Tobago; these are the largest brain corals in the world. Its smooth ridges and wide grooves are home to many species of cleaning goby, depending on region. The polyps only come out to feed at night.

2 GROOVED BRAIN CORAL

Diploria labyrinthiformis

This coral rarely grows to over 120cm (4ft) across. Its intact raised ridges have obvious grooves, but they seem to unravel to form various weird shapes, not necessarily in a brain coral shape. It is fairly common through the region.

3 WHITESTAR SHEET CORAL

Agaricia lamarcki

Sheet corals form large overlapping plates, usually on the outer reef wall, and will be found from the reef crest down to well over 30m (100ft). Common everywhere in the region, this species has obvious lines that follow the form of the outer whorl. The coral is usually creamy ochre, but each small polyp forms a small white star.

4 SCROLL CORAL

Agaricia undata

This sheet coral forms small folds to around 1.8m (5ft 11in). It can grow over other fan corals and is often found on the leading edge of a deep, sloping wall. It occurs in the central to southern Caribbean and prefers well-aerated water.

5 SOLITARY DISC CORAL

Scolymia wellsi

Pale with a raised fleshy ridge, this coral is circular to oval in shape, and small at only 5cm (2in). It is quite difficult to identify without microscopic identification, though note the thin teeth on the rim. It prefers low light conditions and only extends the polyps when feeding at night.

6 SPINY FLOWER CORAL

Mussa angulosa

Colonies of this species can grow to around 60cm (2ft), with individual polyps around 10cm (4in) across. It has a rough texture and a fleshy outer ridge that is fairly convoluted. Its colours range from ochre to green, with a white centre. It is common in all areas of the Caribbean zone, and is typically found in the mid-reef.

7 ATLANTIC MUSHROOM CORAL

Scolmia lacera

This species grows as a single, circular polyp, as large as 15cm (6in) across. It has a variable green and blue striped fleshy margin. It prefers a deeper reef, often on steep walls and amidst other coral outcrops, and is found in most areas of the central Caribbean.

8 KNOBBY CACTUS CORAL

Mycetophyllia alicae

This sheet coral's dome grows to around 50cm (1ft 8in). Its colours vary from grey-brown to cream, and it has small raised ridges and raised dimpled polyps. Preferring deeper water, it occurs on the outer reef walls below 25m (80ft) and is reasonably common in all areas, although it is often mistaken for other cactus corals.

9 ROUGH CACTUS CORAL

Mycetophyllia ferox

Very similar to other cactus corals, this rather small species tends to be flatter and often hangs downwards. Its ridges are longer and it has raised polyps. Most often found in areas of strong water movement, it occurs to around 60cm (2ft) over the entire range.

1
2
3
4
5
6
7
8
9

1 ARTICHOKE CORAL

Scolymia cubensis

With a fairly flat circular disc, this species can grow to 10cm (4in) across. Its flesh may have a wide, irregular, bluish outer ring, with more random markings in the centre.

2 ROBUST IVORY TREE CORAL

Oculina robusta

This coral forms small bushy or tree-like forms from a single base. It is thickly branching with very obvious and raised polyps. It is common in the western Atlantic and particularly around Florida and the Bahamas, but thought to be absent in the western Caribbean. This photograph was taken at Blue Heron Bridge.

Sea Fans

Order: Alcyonacia

GORGONIANS OR SEA FANS are members of the class Octocorallia, characterised by eight tentacles, and do not have a free swimming medusa stage. Their tentacles have tiny feather-like projections. They are very common on shallow reefs, with the Venus Sea Fan especially predominant here. Many species are found on vertical walls. Their fans extend out to collect passing plankton.

3 VENUS SEA FAN

Gorgonia flabellum

The Venus Sea Fan can reach over 1m (3ft 3in). It is very similar to the Common Sea Fan, often having a purplish base, but unlike that species it has small branchlets growing from the sides. It is common on coral planes before the drop-off, and offers shelter to groupers and squirrelfish. It is regarded as abundant throughout the range, but less so in eastern regions.

4 COMMON SEA FAN

Gorgonia ventalina

This species is almost always purple, with a very tight network of flat interconnecting branches. It tends to hug the area where the coral reef drops over onto a mini wall or a true deep wall. It occurs through the entire range and north-east to Bermuda.

5 BUSHY BLACK CORAL

Antipathes caribbeana

This species is a true coral of the class Hexacorallia, but is described here because of its close similarity to a sea fan. The largest and most obvious of the black corals, its branches are brown to black, but can be tinted green or golden colours and can reach over 4m (13ft). Its shape is irregular, resembling a straggly tree, and it is always found on deep walls that drop well below the safe diving depth. Here there is no marine protection available, and it has been harvested indiscriminately for the jewellery and curio trade.

6 GOLDEN SEA SPRAY

Heterogorgia uatumani

A small sea fan, only growing to 30cm (1ft), this distinctive species is golden yellow. It prefers deep water, below 24m (80ft), growing on cliff walls and deeper underhangs and recesses. Most common in Mexico, Cozumel and Cuba, it is rare in other areas of the Caribbean.

7 RED POLYP OCTOCORAL

Swiftia exserta

This small, bush-like gorgonian grows to 50cm (1ft 8in). It is instantly recognisable with its red polyps; the branches are fairly thin and often red or golden yellow. Fairly widespread over the entire range, it is regarded as uncommon.

8 DEEPWATER SEA FAN

Iciligorgia schrammi

This brown sea fan can reach around 120cm (4ft) and many may grow clumped together. It is common in water only 12m (39ft) off Spot Bay on the north shore of Cayman Brac, where the reef wall is very close to shore. It loves the almost constant current that brings its planktonic food.

1 WIDE-MESH SEA FAN

Gorgonia mariae

This species grows to around 30cm (1ft), and grows on a single plane, its pale branches barely interconnecting, if at all. Most common in the central to western Caribbean, it can be found in all depths, but does like clear, clean water with a constant current.

2 BRILLIANT SEA FINGERS

Titanideum frauenfeldii

This species can grow to around 60cm (2ft) but is usually much smaller. Resembling the soft corals of northern waters, it grows as single or sometimes branching slim stalks that rise out of the seabed with a soft, red column and white polyps. It is common in the eastern Caribbean around the east coast of Florida; this photograph is from Blue Heron Bridge.

3 PINNATE SPINY SEA FAN

Muricea pendula

This sea fan grows to around 10–45cm (4–18in), aligned perpendicular to the current. It is brilliant deep red, and is rare in most areas of the Caribbean, except for the far south-east. I photographed this individual in Trinidad in the Bocas Islands, which once linked Trinidad to Venezuela.

Sea Whips

Order: Alcyonacia

THE TRUE SOFT CORALS that you would encounter in other tropical waters such as the Red Sea are absent from the Caribbean. Nevertheless, divers in the Caribbean tend to call all of the following species 'soft corals' because they lack a rigid skeleton.

4 YELLOW SEA WHIP CORAL FAN

Pterogorgia citrina

A small sea whip that only grows to 30cm (1ft), this species has flat branches extending from a short stalk that splits as it extends out into the current. A shallow-water species, it is rarely found deeper than 12m (39ft) and enjoys shallow, clear, clean, well-aerated water. It is very common to abundant in most areas of the Caribbean, and as far north-east as Bermuda.

5 ANGULAR SEA WHIP

Pterogorgia anceps

Looking almost like a terrestrial cactus with straight sides and evenly spaced small polyps all the way up the edges of the blades, this species has many branches and can grow to over 60cm (2ft). It is found in most areas of the Caribbean zone, sometimes very abundantly when conditions are perfect.

6 GIANT SLIT-PORE SEA ROD

Plexaura nutans

This is a tall species of the family Plexauridae, which can grow to over 1.5m (4ft 11in). Its branches are the same thickness as the stem. When the polyps are retracted, the polyp mouth is a small slit, as opposed to the round mouths found on other species. It is common everywhere there is clear, clean water!

7 BLACK SEA ROD

Plexaura homomalia

These polyp colonies grow to around 60cm (2ft). They are congregated into a bushy structure that is generally dark brown, but as always, there are colour variations that make you question identification. The branches are rounded and extend out from a common stalk. It grows in most depths of water.

8 BENT SEA ROD

Eunicea flexuosa

This species grows to 45cm (1ft 6in) and is usually coloured in shades of green and cream. Several thick stalks may extend from a common adhesive body amongst many other corals, becoming flat planes positioned perpendicular to the current. It enjoys shallow waters of less than 3m (10ft), but can also be found deeper in open water bank reefs. It occurs in most of the region.

1
2
3
4
5
6
7
8

1 SEA PLUME

Antillogorgia sp.

The largest of the sea rods, these species' long feathery plumes can be around 2m (6ft 6in) long and may form an irregular long shaggy bush, or consist of just two or three stalks. Sea plumes are common in the entire range, with several very closely allied species that are almost impossible to accurately identify.

2 KNOBBY SEA ROD

Eunicea sp.

The knobby sea rods are somewhat untidy-looking, with quite a thin round stalk and long coral polyps. They can be found in most coral reef systems through the entire range. They are favourite hosts for the Crisscross Tritonia nudibranch (see page 190).

3 BUSHY SEA WHIP

Nicella schmitti

This species grows to a length of 60cm (2ft). It is reddish brown with white polyps. Its whip-like, often split branches extend from a single base stalk. Usually found on reef walls and outer bank reefs deeper than 18m (60ft), it occurs from the Bahamas and Turks and Caicos west through the Caribbean.

4 WIRE CORAL

Cirrhipathes leutkeni

As the name suggests, this species forms long, often spiralling colonies of polyps which extend from a single rod or wire-like base, and can easily reach over 4m (13ft). It is brown or olive green with paler polyps. Found throughout the entire Caribbean range, it is almost always on deeper reef walls.This species plays host to the Wire Coral Shrimp (see page 160).

Hydroids

Order: Anthoathecata

COMMONLY REFERRED TO as 'sea firs', hydroids usually live attached to rocky substrates. Often mistaken for plants or algae, they form delicate fern or feather-like groups and can be quite profuse in certain areas. During sexual reproduction, they produce a tiny free-swimming form called a hydromedusa. Although most hydroids have tiny polyps, they pack powerful stings and contact with them should be avoided.

5 BRANCHING FIRE CORAL

Millepora alcicornis

Yellowish ochre with white tips, this species is hard and calcareous in structure, resembling a true coral. This and other fire corals are well named – the painful sting from their millions of nematocysts will burn unprotected skin and you may require medical attention.

6 BLADE FIRE CORAL

Millepora complanata

Very closely related to the Branching Fire Coral, this ochre-coloured species grows in large flat plates rather than branching fingers. It grows upwards and often becomes quite convoluted in shape. Like its relative, it is armed with serious weaponry.

7 ROSE LACE CORAL

Stylaster roseus

Whilst this colourful pink and white species is regarded as a hydroid, it lacks the more lethal stings of its counterparts, although it can still raise a nasty welt on soft tissue. Thankfully it tends to live in more inaccessible coral recesses, caverns or under rocks in sheltered waters.

8 SOLITARY GORGONIAN HYDROID

Ralpharia gorgoniae

Often overlooked, this species is very like the almost identical tubularian species found in much colder waters. It grows as a single large polyp, attached to the end of a sea plume stalk.

1

2

3

4
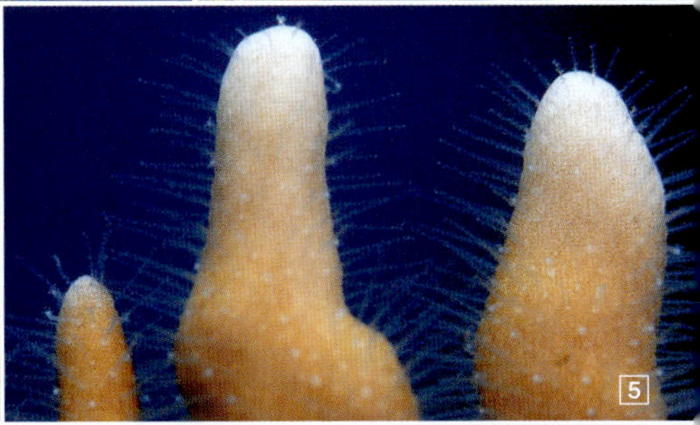
5

6

7

8

1 **BRANCHING HYDROID**

Sertularella diaphana

This hydroid grows to over 10cm (4in). It slightly resembles a feather, with alternating clusters of polyps on slender stalks extending from a thicker central stalk. It is common through the entire range.

2 **UNBRANCHED HYDROID**

Cnidoscyphus marginatus

This species grows as small single stalks, under 10cm (4in) long, which have obvious cream polyps and stretch out from sponges. It is fairly common in the entire range and is found in all depths.

3 **SLENDER FEATHER HYDROID**

Gymnangium speciosum

This hydroid has a slender brown stalk growing to 13cm (5in) with alternating white branchlets of equal length. It is very venomous and should be avoided, but you will encounter it throughout the entire Caribbean range.

4 **CHRISTMAS TREE HYDROID**

Pennaria disticha

Forming small clumps that grow to around 10cm (4in) and have obvious white polyps, this species is found in most areas, on most types of substrate. It prefers clean water with plenty of current.

5 **CARIBBEAN ALGAE HYDROID**

Thyroscyphus ramosus

This hydroid is fairly common everywhere and grows to 13cm (5in). It is yellow with a light brown stalk. It grows in clumps and, as the name suggests, the branches are often covered in algae. It is slightly venomous and is found on shipwrecks, coral rubble and everything else in between!

Jellyfish

Class: Scyphozoa

JELLYFISH ARE GENERALLY cup-shaped and are unattached, free-swimming medusae. Nearly all have a bell or dome with a fringe of nematocyst-bearing tentacles that hang down below. The stinging cells hook and then trap prey, and bear it up to a central mouth. Some tend to spend the daylight hours in deep water and rise up to shallow waters at night, moving either through pulsations of their bell or by beating hair-like cilia. Others stay on the surface and allow the wind to propel them at its whim.

6 **PURPLE or PELAGIC JELLYFISH**

Pelagia noctiluca

This oceanic jellyfish has a large, spotted bell of around 10cm (4in). The spots are actually warts armed with stinging cells. It has 16 lobes, eight sense organs and eight marginal tentacles with four shorter frilled mouth tentacles under the bell. The marginal tentacles can extend over 1m (3ft 3in) long, making them rather hazardous to swimmers during the summer months when this jellyfish can be quite prolific in inshore waters. There is no sessile stage in its reproduction, with the adult releasing juvenile medusae in the autumn.

7 **MOON JELLYFISH**

Aurelia aurita

The Moon Jellyfish is one of the most common of all jellyfish and occurs in all of our oceans and seas. Recognised by the four purplish or blue gonad rings in the top of the bell, it can grow up to 40cm (1ft 4in) in diameter. The bell is surrounded by a ring of small tentacles. It is not harmful to humans, but preys on small fish and planktonic larvae. This jellyfish uses the sun as a compass and forms breeding aggregations in late summer, followed by extensive migration into shallow coastal waters. The sedentary stage is found under rocky overhangs, well aerated by tidal movement, and it releases young medusae into open waters in the spring.

1

2

3

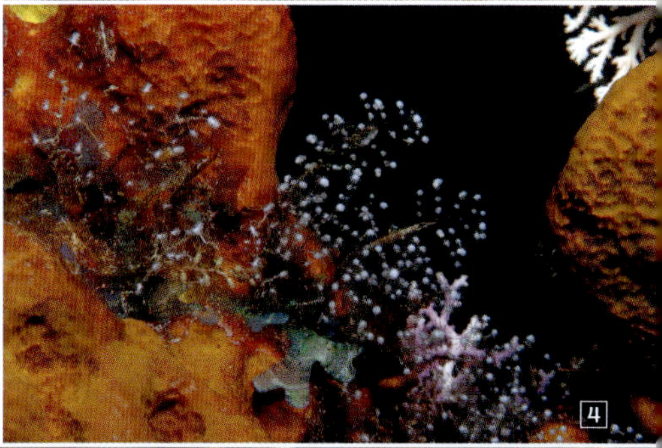
4

5

6

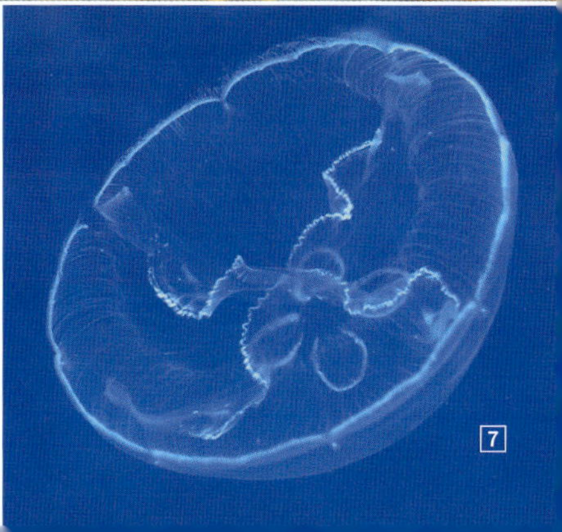
7

1 CANNONBALL JELLY

Stomolophus meleagris

Growing into an almost perfect globe around 25cm (10in) across, this species has quite a short skirt around the edge of the bell, and small tentacles. The skirt is reddish, gradually changing to blueish-green, and the tentacles are blue-white. Juvenile jacks often hang around this jellyfish for protection.

2 UPSIDE-DOWN JELLYFISH

Cassiopea frondosa

Growing to around 13cm (5in), this species swims with the bell on the bottom and the oral tentacles on top. The rim of the bell is often spotted. Feather-like or leaf-shaped tentacles are also present and the jellyfish can pulse its bell for propulsion. Only mildly venomous, it is found in sheltered lagoons and seagrass beds.

3 MANGROVE UPSIDE-DOWN JELLYFISH

Cassiopea xamachana

A larger jellyfish at almost 20cm (8in) across the bell, this species has longer arms than the Upside-down Jellyfish, which are always aimed at the sun to help speed the growth of its symbiotic algae. Its oral tentacles have darker-coloured leaf-like appendages. Found around mangrove lagoons, it tends to rest on the bottom with its oral tentacles above. It is fairly common throughout the Caribbean and Gulf of Mexico.

4 SEA WASP

Carybdea alata

The most venomous of all the Caribbean jellyfish, the Sea Wasp's bell can reach almost 10cm (4in). It is roughly rectangular, with a single tentacle at each corner. It is attracted to lights along the shore at night as well as lights from dive boats, so great care should be taken on night dives. Severe stings will require medical attention, but as a stopgap, pour vinegar over the welts.

5 BLUE BUTTON

Porpita porpita

The Blue Button is very distinctive. A jellyfish-like colonial animal similar to the Portuguese Man-of-war, it has a round float about 2.5cm (1in) across, and a hydroid colony below, which resembles the tentacles of a jellyfish. It floats on the surface and is just as at home in the Mediterranean as in the Caribbean.

6 PORTUGUESE MAN-OF-WAR

Physalia physalis

This is a very distinctive siphonophore with an obvious sail-like bladder or pneumatophore, coloured blue, pink or mauve, which sits above the surface of the water. It can grow as large as 9–30cm (3½–12in) long and be as tall as 15cm (6in). The colony of zooids which it supports has stinging tentacles that can grow to 10–30m (33–100ft) long. It moves by wind power and may have very long, highly toxic tentacles. During storms, literally hundreds can be washed up on the shore and they must be avoided at all costs.

7 SEA THIMBLE

Linuche unguiculata

A small jellyfish, the Sea Thimble's bell only reaches about 2cm (¾in). There is a ring of a few small tentacles around the edge of the dome. The larvae are microscopic and present in the water column at the same time as the adults; exposure to them can leave an irritating rash for over a week. It is found around coastal waters on many Caribbean islands, often in staggeringly huge numbers in the springtime.

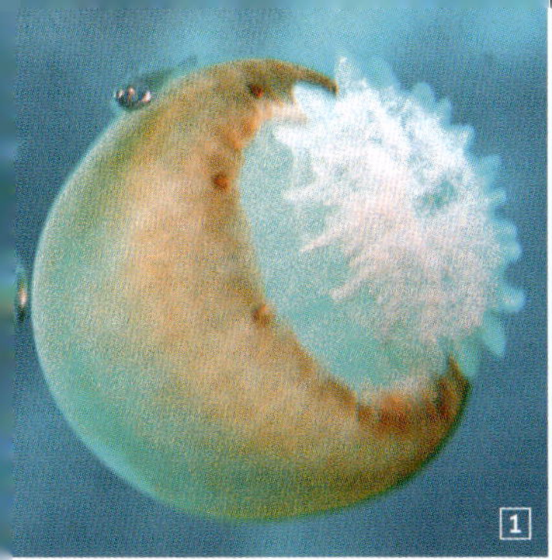
1

2

3

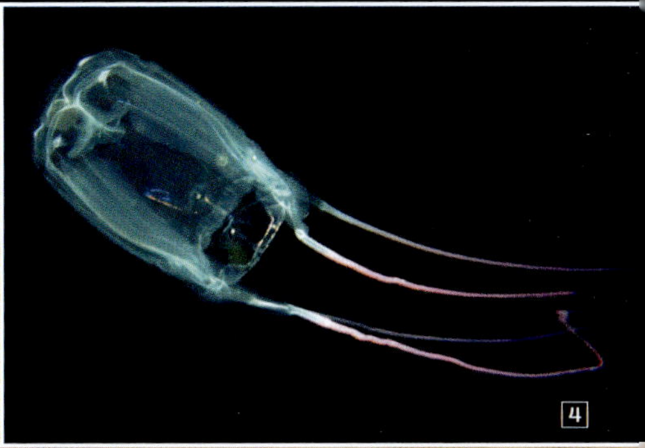
4

5

6

7

CRUSTACEA

Subphylum: Crustacea

THIS LARGE GROUP has more than 30,000 members worldwide, most of them marine-dwelling. In general, crustaceans have segmented bodies with a head, thorax and abdomen, and are heavily protected by calcareous outer body armour. Most divers see the obvious decapod species, which have ten legs that are adapted variously for walking, feeding, swimming, defence, food capture, respiration and even the carrying of eggs, depending on the species.

Barnacles

Subclass: Cirripedia

1 GOOSE BARNACLE

Lepas anserifera

The Goose Barnacle is around 4cm (1½in) long. It has a long, leathery foot and a strong, calcified, protective head where the 'catching mitt' is kept. There is a stalk between the head and the foot. It attaches itself onto floating bits of debris, ships' keels and piers, so can travel widely despite being sessile. It is tolerant of poor water conditions.

2 SESSILE BARNACLE

Balanus sp.

These barnacles are permanently attached to a number of different surfaces, including natural substrates but also piers, jetties and slipways. They have a protective exoskeleton of overlapping shell-like plates and feed through the top with fan-like cirri which extend to capture any planktonic particles that float by. They can grow to over 2.5cm (1in).

Hermit Crabs

Superfamily: Paguroidea

HERMIT CRABS INHABIT discarded mollusc shells, or other portable shelters that can protect their soft abdomen. They carry their mobile homes around with them, upgrading to larger shells as they grow. When in danger they retreat into the shell, either completely or form a hard protective barrier where the mollusc's operculum once was.

3 GIANT HERMIT CRAB

Petrochinus diogenes

This species, the largest of the Caribbean hermit crabs, grows to over 20cm (8in) across, and has knobbly reddish claws with red and white striped antennae. As it grows older and larger, suitable shells become more difficult to find, so it often inhabits a much larger shell when young and gradually grows into it! It is found in most locations at all depths.

4 STAREYE HERMIT

Dardanus venosus

This hermit crab can grow over 10cm (4in). It has red and white stripes and pale blue eyes. It also has the privilege of carrying around the Hitchhiking Anemone (see page 124), attached to its adopted home. It occurs in a wide variety of habitats from seagrasses to coral reefs, but is most at home on the seabed amongst coral rubble.

5 RED-BANDED HERMIT

Paguristes erythrops

Very similar to the Stareye Hermit, this species only grows to around 8cm (3¼ in). The red and white stripes on its claws and legs are much more obvious and there may be purple tips to the tubercules on its claws. It also has long stalked blue eyes. It is not as common as the Stareye in the region and may be misidentified.

6 WHITE SPECKLED HERMIT CRAB

Paguristes punticeps

Growing to around 13cm (5in) across, this hermit crab is dark reddish brown and covered in white spots with white bands at each of the leg and claw joints, and has a more flattened body than other hermit crabs. It is found in most areas of the Caribbean zone.

1
2
3
4
5
6

1 RED REEF HERMIT CRAB

Paguristes cadenati

This is the most common reef hermit crab in the region. It grows to around 2.5cm (1in) and has bright red legs and claws and yellow-stalked green eyes. It occurs mainly on the reef and coral rubble areas, where it forages relentlessly.

2 BLUE-EYE HERMIT CRAB

Paguristes sericeus

This species grows to around 5cm (2in). It is mainly orange-red, with long red stalks to its bright blue eyes. Its body is flattened so it prefers shells with a narrow opening, such as cone shells or small winkles. It is fairly common in the entire region.

3 RETICULATED HERMIT CRAB

Iridopagurus reticulatus

This is a tiny hermit crab, only around 1cm (⅓in). It inhabits tiny winkle shells by preference, but is often shell-less. It is white with reticulate markings on the legs and claws, and carries its black eggs just inside the shell opening above its head. It occurs on sandy seabeds and is extremely speedy across the sand when necessary.

4 RED-STRIPE HERMIT

Phimochirus hothusi

This is a small crab, only growing to 2.5cm (1in). Its name is a misnomer as all of the specimens seen and photographed have distinctive brown stripes down all of the legs and claws. It can be found on soft corals, sea whips and other coral reef surfaces. The right claw is much larger than the left.

5 ORANGECLAW HERMIT

Calcinus tibicen

This crab only grows to around 2.5cm (1in). Its claws are equal-sized and it has long, thin legs in a variety of colours, from orange to red and brown; the body is paler. It has orange eye-stalks and antennae, while the eyes themselves are black with white surrounds. It is found in most habitats in most depths.

True Crabs

Infraorder: Brachyura

True crabs are decapods which often have their legs hidden under the shell. The front pair of legs have adapted into powerful claws. The greatly reduced tail has all but disappeared and is held as a bony flap under the body. Some species are more mobile than others, while some have flattened rear legs which they use for swimming in a sideways direction.

6 CHANNEL CLINGING CRAB

Mithrax spimospssimus

The largest of the coral reef crabs to be found in the region, this crab's carapace can be as large as 20cm (8in) across. It has eight long, strong legs and two very large, smooth pincers, the ends of which are lighter in colour than the rest of the brownish-red body. It is nocturnal and its name comes from its habit of climbing up sea fans and other corals. It occurs in all areas.

7 BANDED CLINGING CRAB or ANEMONE CRAB

Mithrax cinctimanus

Almost always found in association with either the Sun Anemone or the Giant Anemone (see page 120), this crab enjoys the protective embrace of the anemone's stinging tentacles. It is common throughout the region but is often overlooked due to its small size – only 2cm (¾in). It has smooth, blunt claws and is banded in cream, brown and reddish over its knobbly body.

8 RED-RIDGED CLINGING CRAB

Mithrax forceps

A similar size to the Banded Clinging Crab, this species has a deeply ridged carapace. The claws bear sharp nodules. It is a nocturnal feeder and occurs in all habitats, from Bermuda to Brazil.

1
2
3
4
5
6
7
8

1 HEART URCHIN PEA CRAB

Dissodactylus primitivus

A tiny, round, white crab, this species lives amongst the hairs of the large Heart Urchin. It is difficult to find due to its size at only 1cm (⅓in), but does stand out thanks to its colour. It is covered in minute white hairs and may have an intricate pattern over its back.

2 NIMBLE SPRAY CRAB

Percnon gibbesi

This crab grows to around 5cm (2in). It is found in very shallow water at the surf line. It often shelters under sea urchins but will flit between the water's edge to the dry nearshore and back into the water. It is very nimble and quick to get away if danger threatens.

3 YELLOWLINE ARROW CRAB

Stenorhynchus seticornis

A very distinctive crab, this species has a triangular body, long spindly legs and a pointed rostrum. It is orange with long stripes down each limb. The claws are usually tipped in light purple. It is an active feeder on the coral reef at night.

4 CRYPTIC TEARDROP CRAB

Pelia mutica

This small crab grows to 2cm (¾in) and has bluish claws. Its hairy legs are covered in small bits of sponge, tunicates and hydroids. It climbs out onto various corals at night, and although it is small, it is easily seen as it is fairly common everywhere.

5 ANTILLES SPONGE CRAB

Cryptodromiopsis antillensis

This decorator crab is quite robust and grows to 5cm (2in) across the carapace. Its eyes are on long pale stalks. It keeps a trimmed piece of sponge or two on its back, held in place by the two pairs of modified rear legs. It is nocturnal and difficult to spot, but if you ever see a sponge 'walking', then this is probably what is underneath it!

6 RED EYE SPONGE CRAB

Dromia erythropus

Brown and completely covered in tiny hairs, this decorator crab measures about 7.5cm (3in) and, as the name suggests, has red eyes and red tips to its claws. A shy crab, it only comes out at night and holds its living sponge hat with two pairs of short legs.

7 BATWING CORAL CRAB

Carpilius corallinus

Looking similar to the British Edible Crab (*Cancer pagurus*), this crab grows to over 10cm (4in) and is reddish brown with a paler underside and an intricate pattern of spots all over the carapace. It is commercially harvested and therefore may be rare in some locations, but it is generally a common nighttime reef crab.

8 NODOSE RUBBLE CRAB

Paractaea rufopunctata

At only 2.5cm (1in), this is a tiny crab, with a hairy carapace and claws that are covered in large, raised orange bumps. It prefers coral rubble areas and spur and groove reefs, and is found from the Carolinas south to Brazil and all through the rest of the range.

9 GHOST or SAND CRAB

***Ocypodinae* spp.**

This semi-terrestrial species is found around most sandy shorelines throughout the entire region. It makes burrows in the sand above the low water line and scavenges around the beach in detritus. Though quite shy, it will defend its territory. Pale cream in colour with black eyes, it grows to around 35mm (1.4in).

1

2

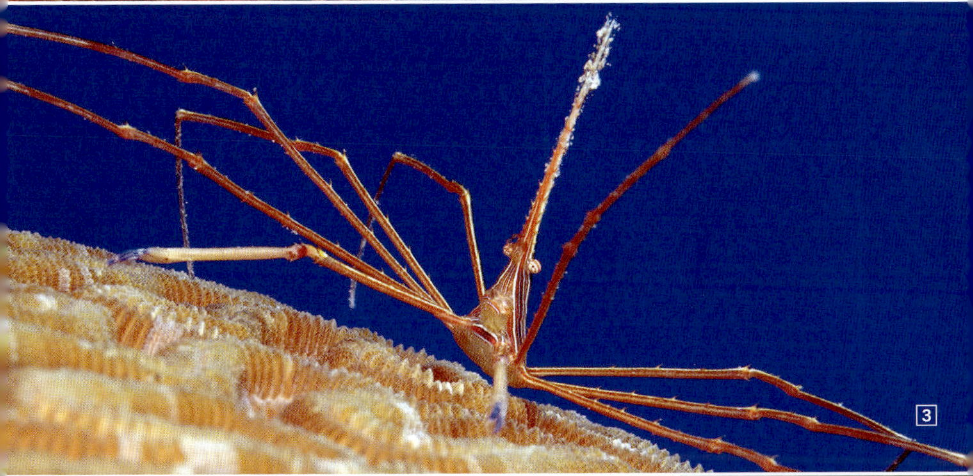

3

4

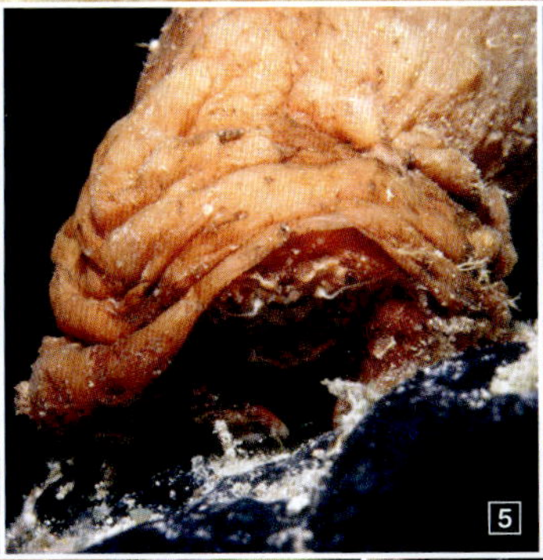

5

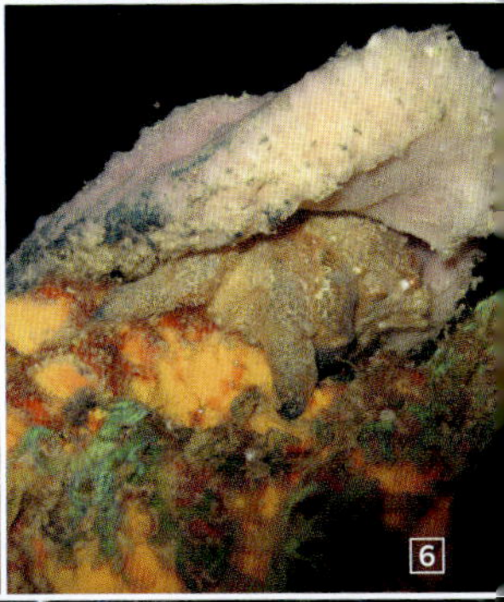

6

7

8

9

1 FLAME BOX CRAB

Calappa flammea

This curious crab grows to 10cm (4in) across its back. It has small eyes on stalks, and a pale body with red flame-like markings all over the carapace and side of its claws. Members of the genus *Calappa* have very robust claws which they can fold into the body, making them almost impregnable to attack. This species occurs from Bermuda down into the Gulf of Mexico.

2 SHAMEFACE HEART CRAB

Cryptosoma bairdii

This crab grows to around 7.5cm (3in) and has roughly textured flat claws that it can fold flat against the front of its body. It is nocturnal and usually hides below the sand during the day, with only its tiny stalked eyes visible. It is very widespread from the Galapagos and Gulf of California to the western Atlantic and central Caribbean.

3 FLORIDA DECORATOR CRAB

Macrocoeloma camptocerum

This green or brown spider crab grows to around 5cm (2in). It tends to hide amidst the algae and seagrass beds, and its spiny shell often has algae and bits of live sponge attached for camouflage. It is quite rarely seen due to its camouflage and nocturnal habits, but is found through the Caribbean and Gulf of Mexico.

4 CARIBBEAN REDHAIR SWIMMING CRAB

Achelous ordwayi

Tan, orange and red in colour, this is a small swimming crab growing to around 5cm (2in). Quite an aggressive crab, it has three prominent spikes on its claws which are always at the ready in a defensive posture. It ranges from Bermuda to Brazil and is common along Florida's east coast, particularly at the Blue Heron Bridge.

5 BLOTCHED SWIMMING CRAB

Portunus spinimanus

Similarly shaped to the previous species but larger, growing to 7.5cm (3in), this swimming crab is dull brown, blotched with brown markings. The rear legs are paddle-shaped for swimming which it uses to great effect when you get too close and ignore its aggressive posture. It is common in most areas of the western Atlantic and into Florida waters.

6 OCELLATE SWIMMING CRAB

Portunus sebae

This very aggressive swimming crab grows to 10cm (4in) and is tan to red in colour with two very obvious ocellated spots to the rear of the carapace (almost like large eyes). It spreads its claws very wide, making it look even larger.

7 SHORT FINGER NECK CRAB

Coryrhynchus sidneyi

This is one of a number of similar-looking decorator spider crabs. They grow to around 5cm (2in) and have triangular bodies with very long legs and pincers. Most cover themselves in sponges, hydroids and algae, and when hiding amongst the same species that they have attached to themselves, they are almost invisible. They are common everywhere, but rarely observed.

8 CURVED-NOSE NECK CRAB

Anisonotus curvirostris

Growing to around 6cm (2½in), this decorator crab is more commonly found in larger tube sponges and usually completely covered in hydroids. Cream to yellow/pink in colour, it may also have orange spots around its up-curved rostrum. It is relatively common in the central Caribbean.

1

2

3

4

5

6

7

8

1 **HORSESHOE CRAB**

Limulus polyphemus

Horseshoe crabs are actually closer cousins to spiders and belong to the class Merostomata. This species is grey to brown and can grow to almost 60cm (2ft). It is common in Florida waters and all the way up the Delaware coast where there are mass migrations. It is important in the food chain, the adults being eaten by turtles and the young by seabirds. Horseshoe crabs are often described as living fossils, being little changed from their fossil relatives that lived in the Paleozoic era around 445 million years ago.

Lobsters

Family: Nephropidae

COMMERCIALLY IMPORTANT in many locations, lobsters are nocturnal bottom-dwellers and have strong legs to help them clamber over the reef. Most Caribbean species have no obvious claws (unlike the cold-water lobsters).

2 **CARIBBEAN SPINY LOBSTER**

Panulirus argus

This is the largest lobster in the range, sometimes reaching 60cm (2ft) long. It has very large and sensitive antennae, and the body is variously brown, red and ochre. The carapace has very sharp points. It can often be found in large groups under overhangs. Known to mass migrate into shallower waters to spawn, it reverses the journey into deeper waters when winter storms are imminent. It is the most common lobster in the entire range.

3 **GUINEA CHICK or SPOTTED LOBSTER**

Panulirus guttatus

Much smaller in size than the Caribbean Spiny Lobster at around 30cm (1ft), this species is characterised by its abundant white spots, over a brownish/purple body. I first encountered it in large numbers at Flatt's Bridge in Bermuda.

4 **RED BANDED LOBSTER**

Justitia longimanus

Quite rare in the entire area, this lobster only forages at night. It grows to about 20cm (8in) and is red, with cream at the hinges of the claws and legs, and sometimes various white bands and spots. I have only ever found one once, on the island of Dominica.

5 **SPANISH LOBSTER or SHOVEL-NOSE**

Scyllarides aequinoctialis

A large species, growing to around 30cm (1ft), this lobster's antennae are formed by large overlapping plates. Its legs are strong and abdomen quite short, but it has a typical lobster shape. It can swim backwards swiftly if in danger.

6 **SCULPTURED SLIPPER LOBSTER**

Parribacus antarcticus

Much smaller at around 20cm (8in), this is another species with lobed antennae. The frontal lobe's edges are covered in small spines and bristly hairs. Often found in pairs on the coral reef at night, it is fairly common across the entire zone.

7 **CARIBBEAN SQUAT LOBSTER**

Anomoeomunida caribensis

By far the smallest member of the group, this squat lobster is perhaps 2.5cm (1in) in total length, with disproportionately long, thin claws. It is pale cream to orange-red in colour. Usually found amongst coral rubble, it is often overlooked due to its small size, and is widespread throughout the entire range.

Shrimps

Orders: Decapoda, Stomatopoda

THESE DECAPODS are characterised by their long antennae and laterally compressed bodies. The abdomen is usually quite large, to allow for speedy movement in reverse. Shrimps vary greatly in colour and habitat. The most commonly seen are the 'cleaning shrimps', which remove parasites and decayed flesh from many different species of fish. Some other shrimps live in burrows, and most species are active predators at night.

1 SCALY-TAILED MANTIS

Lysiosquilla scabricauda

This is a 'spearing' mantis shrimp, capable of inflicting a painful wound should you care to agitate it! It grows to 15cm (6in) and is quite robust, cream in colour with brownish stripes and bands, and with large, stalked, double-lobed eyes. Widely distributed throughout the entire range, it lives in well-formed circular burrows on flat sandy seabeds.

2 SWOLLEN-CLAW MANTIS SHRIMP

Gonodactylus oerstedii

This is a 'smashing' mantis, using its claws to stun its prey with the force of its blows. It grows to 5cm (2in). It is an active predator and forages over all areas of the reef, often living under stones and in recesses. It is not often seen, but is occasionally found in most areas.

3 CILIATED FALSE SQUILLA

Pseudosquilla ciliate

Also growing to around 5cm (2in), this species is generally green with small, stalked, single-lobed eyes. Usually found in beds of seagrass, it makes U-shaped burrows under corals and coral rubble. It is fairly common in most of the Caribbean zone.

4 BANDED CORAL SHRIMP

Stenopus hispidus

This is a distinctive species with a white body and claws, banded with dark red and black blotches. It waves its six long white antennae to advertise its services for fish requiring a special clean-up! Very common throughout the entire range, they are circumtropical and you will find them in all tropical, warm-water oceans around the world.

5 GOLDEN CORAL SHRIMP

Stenopus scutellatus

Very similar to the Banded Coral Shrimp, this is a smaller cleaner shrimp, coloured golden yellow with darker orange and white bars across its claws and rear of the abdomen. Often associated with sponges and deeper recesses, it is quite wary and quickly retreats if you approach it.

6 PEPPERMINT SHRIMP

Lysmata wurdemanni

This grows to almost 5cm (2in) long and is pale orange or pink with red stripes over the entire body. This is another 'cleaner', although it is rarely observed in this role as its cleaning station is usually well hidden.

7 SCARLET LADY

Lysmata grabhami

This is another very distinctive little shrimp, growing to 5cm (2in). It has six long white antennae and its two front legs are also white. The body is golden red and it has a prominent white stripe down the entire length of the body. It is an active 'cleaner' on many species of fish. I have had my teeth cleaned by one, while diving on Cayman Brac!

8 SQUAT ANEMONE SHRIMP

Thor dicaprio

Only reaching 2cm (¾in), this species is greenish yellow with prominent white eyes, a white thorax and banded white abdomen with small white spots, often all lined with purple. It lives within the embrace of several different anemone species, and is common throughout the entire region.

1
2
3
4
5
6
7
8

1 SEA PLUME SHRIMP

Neopontonaides chacei

This tiny yet robust shrimp, less than 1cm (⅓in) long, has a flattened rostrum with a prominent point. It has protruding eyes and tiny legs ending in hooks, with which it traverses the soft stalks of sea plumes. I have only ever seen it at night.

2 ROUGHBACK SHRIMP

Trachycaris rugosa

This species grows to around 2½cm (1in). It has a distinct hinged back and is often covered in all manner of algae bits to add to its camouflage. It occurs throughout the Caribbean though is rarely seen due to its small size and camouflage.

3 BERMUDA SPONGE SHRIMP

Holthuisaeus bermudensis

This shrimp is pale pinkish-tan with black eyes, and likes to inhabit small sponges. It is quite shy and will retreat into the sponge if approached. It is rare in the entire range, but is certainly found from Bermuda to Brazil and as far west as the Cayman Islands, Cuba and even Panama.

4 PEDERSON'S CLEANER SHRIMP

Ancylomenes pedersoni

This is a very active cleaner shrimp, which waves its long white antennae to attract customers. It has a transparent body with long blue-purple chelipeds; the back is similarly coloured. It is usually found in the vicinity of Corkscrew Anemones (see page 122) where the reef meets the seabed. This is by far the most common shrimp throughout the entire range of the western Atlantic and greater Caribbean zone.

5 SPOTTED CLEANER SHRIMP

Periclimenes yucatanicus

One of the most colourful of all the Caribbean shrimps, the Spotted Cleaner Shrimp has a symbiotic relationship with a few anemones, but principally the Giant Anemone. Stout of body, up to 2.5cm (1in) long and nearly transparent, it has pinkish, well-defined markings with long, blue or red pincers and blue-striped legs. It has similarly coloured markings on each of its tail plates. Seemingly rather docile when near its anemone protection, it is able to dodge very rapidly out of the way should danger threaten. It can also be found around seagrass beds, where its bright colours are a very rare delight for divers. It is almost identical to *P. amethysteus* in the Mediterranean, and *P. sagittifer* in southern British waters.

6 SUN ANEMONE SHRIMP

Periclimenes rathbunae

Often commonly associated with the Branching Anemone (see page 124) and most commonly with the Sun Anemone (see page 120), this is a fairly translucent shrimp that grows to 2½cm (1in). It is covered in small ochre and white spots, stripes and intricate markings. It can be found in most areas.

7 SQUAT URCHIN SHRIMP

Gnathophylloides mineri

Almost impossible to find, unless you examine each and every Variegated Urchin (see page 168), this shrimp is very small, perhaps only 1cm (⅓in). It has black, white and grey stripes down the body. It rests its abdomen within the urchin's tentacles, and feeds on detritus and plankton. This shrimp ranges from Bermuda through the entire range.

8 RED NIGHT SHRIMP

Cinetorhyrehus manningi complex

There is disagreement over the classification of this shrimp – it may have up to six subspecies, some of which are sometimes treated as separate species. It is nocturnal and generally red, but some may have a lighter thorax, others exhibit red spots, others have white spots on a red background, and still others have a banded abdomen! Its reflective green eyes at night alert us to its presence on the reef. It is common everywhere.

1
2
3
4
5
6
7
8

1 VELVET SHRIMP

Matapenaeopsis goodie

Growing to around 5cm (2in), this is a common shrimp with a transparent body, dotted with irregular red markings. It has large reflective green eyes, and the first three sets of legs have pincers on them. It lives in lagoons and around seagrass beds.

2 BROWN SHRIMP

Penaeus aztecus

This shrimp is characterised by the ridge of forward-facing spikes that line the top of the thorax. It is brown and tan with bands and spots, and has stalked eyes. It is found principally around seagrass beds. Fairly common throughout the range and also found in the eastern Atlantic and Mediterranean, it is commercially harvested in some areas.

3 SCARLET PISTOL SHRIMP

Alpheus armatus

This species lives under the tentacles of the Corkscrew Anemone and is characterised by the green-gold spots on the claws and at the end of the rostrum. The 'snap' of its claws is easily heard underwater if the shrimp is agitated!

4 SPOTLESS SNAPPING SHRIMP

Alpheus immaculatus

This species resembles the Scarlet Pistol Shrimp but has no spots. It also appears to be much more mobile, and can be found wandering about amongst the coral rubble. It also likes the protective embrace of an anemone's tentacles.

5 BURROWING PARTNER SHRIMP or SAND SNAPPING SHRIMP

Alpheus floridanus

This shrimp can grow to around 5cm (2in). Like so many of its tropical counterparts, it has a symbiotic relationship with a fishy partner, the Orange-spotted Shrimpgoby (see page 68). The shrimp excavates a burrow and, should danger threaten, the goby retreats into the burrow with the shrimp. It is widespread from Bermuda to Brazil and through the entire range.

6 RED-ORANGE GHOST SHRIMP

Corallianassa longiventris

A curious cross between a snapping shrimp and a mantis shrimp, this species can grow to 10cm (4in) long. Its claws are pointed and always open. It lives in circular burrows in soft sand and resembles a trapdoor spider! This was found at Blue Heron Bridge in Florida.

7 PINK GHOST SHRIMP

Callichirus major

Growing to over 5cm (2in), this elongated species is translucent pink. It can be quite common locally, occurring along the entire western Atlantic in fairly shallow water of only around 3m (10ft). I photographed this fellow in that depth off Anse Chastanet in St. Lucia during a night dive.

8 RED CUBAN CAVE SHRIMP

Barbouria cubensis

A brilliant red colour, this quite robust shrimp is unmistakable. It has a round, plump thorax, black eyes and widely spaced antennae. The tapered abdomen ends in a wide, five-lobed fan tail. This is a cave shrimp, but on Cayman Brac when tides and sun work together, many hundreds can appear in small inland pools to eat algae in the shallow water.

8

1
2
3
4
5
6
7
8

1 WIRE CORAL SHRIMP

Pseudopontonaides principis

Very small at only 1cm (⅓in), this species has much longer claws than other wire coral shrimps, and tends to be the same colour as its host coral (green or brown). Difficult to find and considered quite rare in the Caribbean, it had not been officially recorded in the Bahamas when I photographed this specimen off San Salvador in the southern Bahamas.

2 ARROW SHRIMP

Tozeuma carolinensis

Growing to 4cm (1½in), this elongated shrimp has a humped back near its rear. The long snout tapers to a sharp point and is slightly upturned. Its colour changes according to its habitat; it is most often found amongst algae and seagrasses.The name suggests that it can be much further spread throughout the range of the western Atlantic.

Isopods, Sea Spider and Skeleton Shrimp

Taxa: Isopoda, Cymothidae, Nymphonoidea, Caprellidae

3 GREENSCALE ISOPOD

***Sphaeromatidae* sp.**

This species is very small at around ¾cm (¼in). It is in the marine pillbug family and is surprisingly common in the entire region. It eats detritus and is found around most shores, on many different substrates.

4 PARASITIC ISOPOD

Anilocra myripristis

Despite the name, this small crustacean is not parasitic at all, but hooks onto a fish host for the ride! It is fairly common throughout the range and uses any number of different fish species as hosts. It benignly scavenges fish scraps and plankton and it would appear that the host provides additional protection as well as transport.

5 TUBE-DWELLING ISOPOD

Pagurotanais largoensis

This species is quite shrimp-like in appearance. You will notice that the largest cheliped has a distinct hook on the end of the larger claw. It is characterised by its tube-living habit and is very rare indeed, never known from the Cayman Islands until this photo was taken! It has been recorded in Cuba.

6 CARIBBEAN SEA SPIDER

***Anoplodactulus* sp.**

This pycongonid sea spider has a very small body and four pairs of legs. Very spider-like in appearance, it is blue-green with green and yellow markings and only grows to 1cm (⅓in). It is quite rare overall; this specimen was found in the Cayman Islands.

7 SKELETON SHRIMP

***Caprella* sp.**

Skeleton shrimps are amphipods that only grow to around 2½cm (1in) and are recognisable by their elongated, three-sectioned bodies. The perion or thorax is the longest segment. The cephalon or head has two pairs of antennae, one pair longer than the other. It tends to hang out on various hydroids and bryozoans and is found in all areas.

4

1
2
3
4
5
6
7

ECHINODERMS

Phylum: Echinodermata

THIS IMPORTANT PHYLUM is entirely marine and includes many family groups. Although highly varied in form, they have certain common characteristics, such as calcareous plates, a water vascular system and tiny tube feet for locomotion and capturing food. Many are well protected, most obviously the sea urchins with their covering of sharp spines.

Brittlestars and Basket Stars

Class: Ophiuroidea

BRITTLESTARS, AS THE NAME suggests, have very fragile, long thin arms which radiate from a circular, central disc. They can undulate their arms to allow them some independent movement, otherwise they can walk on jointed legs called cirri. Basket stars have five arms at the base that branch repeatedly to form a huge net of multi-jointed complex arms, sometimes over 1m (3ft 3in) across, that they spread out into the current at night. They are very sensitive to light; during the day, they roll up into a tight ball and hide in crevices.

1 SPONGE BRITTLESTAR

Ophiothrix suensonii

The most commonly seen of the Caribbean brittlestars, this species has a distinctive black stripe down each of its five arms, which grow over 5cm (2in). Found throughout the range, it enjoys a number of habitats including amongst many sponge species, and can be seen from Bermuda to Brazil.

2 BLUNT SPINED BRITTLESTAR

Ophiocoma echinata

This robust brittlestar is quite large, its legs growing up to 15cm (6in). It is black-brown and grey with short blunt spines along all of its arms. It occurs in coral rubble habitats through the entire region.

3 RUBY BRITTLESTAR

Ophioderma rubicundum

Most often found in deeper areas of the reef, this brilliant red brittlestar is unmistakable with its banded arms growing to around 10cm (4in). It prefers hiding away amongst various sponges, but can be found on sea fans and corals later at night.

4 RETICULATED BRITTLESTAR

Ophionereis reticulata

This brittlestar has quite long legs that bear a reticulated pattern of fine dark bands along their entire length. It lives amidst coral rubble and hides under these rocks during the day, coming out to forage amongst the detritus at night. It is common in all areas, but rarely seen.

5 GIANT BASKET STAR

Astrophyton muricatum

This species is quite spectacular at night, when it spreads its metre-wide net to catch plankton. It is common in the entire range and can also be found from North Carolina to Brazil.

5

6 SEA ROD BASKET STAR

Schizostella bifurcata

Usually found tightly curled up around the stems of various sea rods, this is a smallish species, its arms growing to 5cm (2in) long. It is brownish-purple with heavily ribbed legs that only branch a couple of times. It is most common in the central Caribbean.

1
2
3
4
5
6

Crinoids

Order: Crinoidea

CRINOIDS OR FEATHERSTARS also have long fragile arms, but here the similarity to other starfish ends. Often referred to as 'living fossils', they are seemingly unchanged for millennia. Featherstars have more than ten arms, with long spines along the entire length, which they sweep through the current for particles of food.

1 GOLDEN CRINOID

Davidaster rubiginosa

With 20 golden-yellow arms that can grow to 25cm (10in), this crinoid is found on all types of coral reef, where its body is usually hidden in the corals, its arms waving in the current to catch plankton. It is quite common in the central and western Caribbean.

2 SWIMMING CRINOID

Analcidometra amata

This common crinoid has ten white arms banded in reddish-brown broad stripes. Its name comes from the way it can manipulate its arms in undulations to allow it to 'swim', should there be any threat. Usually found in depths greater than 12m (39ft), it is seen on all types of corals, from sea fans to plumes and sea rods.

3 BLACK AND WHITE CRINOID

Nemaster grandis

This is a very obvious large crinoid with around 40 black arms that have white tips to each of the side branches. It occurs amongst various sponges, soft and hard corals and can be seen at any time of the day. It is most common in the southern Caribbean.

4 BEADED CRINOID

Davidaster discoideus

This crinoid has 20 arms with pale grey side branches that appear to be beaded, and have black tips. It usually hides away with only its arms showing, but appears to be an active feeder at all times. Usually found below 18m (60ft), it is relatively uncommon, but has a wider distribution than some other Caribbean crinoids.

5 WEST INDIAN SEA LILY

Cenocrinus asterias

Sessile crinoids are known as sea lilies. This species has a long rigid stalk and a cylindrical fan of more than 60 flexible arms that form an umbrella, reaching around 60cm (2ft) off the seabed. It is found below 90m (300ft) in areas of strong current. First seen from the Grand Cayman Atlantis Submersible at around 223m (730ft), growing near the wreck of the Kirk Pride, it is now observed on most deep dives.

Starfish or Sea Stars

Class: Asteroidea

STARFISH USUALLY HAVE five arms, but some species have more. Usually radially symmetrical, they have triangular arms, and the mouth is located on the underside at the centre of the body, while the anus is on top in the centre. They move with water hydraulics and have small tube feet with suction discs on the end.

6 COMMON COMET STAR

Linchia guildingii

This starfish is orange-tan or yellow with five to seven arms, often of unequal lengths. It has randomly spaced blunt nodules all over the body. Found on all reef types, it is reasonably common in the Florida area and Gulf of Mexico.

7 CUSHION SEA STAR

Oreaster reticulatus

This starfish is usually around 30cm (1ft) across,with five short, blunt arms. It is generally reddish-brown, with raised nodules over the back that form intricate patterns. It can be found in large numbers in some areas.

8 BANDED SEA STAR

Luidia altermata

This starfish is multicoloured in shades of tan, brown, grey and black, and has five longish and rather limp blunt arms. It prefers to live under soft sand and mud, foraging at night. It is widespread but rarely seen due to its behaviour, and is found as far south as Argentina.

1
2
3
4
5
6
7
8

1 **TWO-SPINED or BURROWING SEA STAR**

Astropecten duplicatus

This is a true burrowing sea star and like other members of the globally distributed genus *Astropecten,* it has five triangular arms, the sides of which are armed with sharp spines. Rarely seen because it tends to hide away during the day, it is found throughout the entire range.

2 **ORANGE-RIDGED SEA STAR**

Echinaster spinulosus

This species rarely grows over 15cm (6in) across. It is yellowish-orange or reddish-brown. The arms are upturned at the ends and are slightly darker than the rest of the body. It is found on all types of habitat and in most depths, most commonly around the Florida coast and northern Bahamas.

3 **CONICAL SPINED SEA STAR**

Echinaster sentus

This species has somewhat 'floppy', equal-length arms, covered in rounded spines and tubercles. The tips of the arms are rounded. It inhabits algae-covered rubble areas rich in sediment, such as around mangrove beds and estuaries.

4 **BLUNT-ARMED SEA STAR**

Asterina folium

The smallest of the sea stars, this (usually) yellow or tan species only grows to around 2½cm (1in) across. It has short blunt arms which are rounded at the tips and of equal size. A rare sea star, it usually hides under algae-covered rocks and in seagrass beds, and is found around Florida and much of the western Atlantic.

5 **WHITE SEA STAR**

Coscinasterias tenuispina

This species generally has seven arms, but can have as many as 12, often of varying lengths. It has a rough texture and numerous short spines and reproduces asexually by 'fission', where it literally tears itself in two and grows another set of arms to form a new animal!

Sea Urchins

Class: Echinoidea

SEA URCHINS HAVE a hard rounded skeleton of ten calcareous plates, which are covered in protective spines. The mouth is directly underneath and the bony mouth parts form an intricate structure called 'Aristotle's Lantern'. The spines vary from soft and hair-like, to thick and blunt, to long delicate needles with pin-sharp tips that can easily penetrate human skin.

6 **MAGNIFICENT URCHIN**

Astropyga magnifica

By far the most gaudily coloured of all the Caribbean sea urchins, this species is pinkish-red with sections of spines grouped together in loose clumps. The spines are banded in red and their bases have minute iridescent blue spots. It is quite a rare species, often found only in deep water, but I find it regularly off the east end of Cayman Brac in under 12m (39ft).

7 **JEWEL URCHIN**

Lytechinus williamsi

This brightly coloured species grows to around 5cm (2in) in diameter. The test (empty shell) is creamy white. The living body is also cream, but with red stripes between the segments. It has very long tentacles, greenish spines and the tube feet end in purple bulbs or balls. It is rare but present in many locations of the Caribbean.

8 **LONG-SPINED URCHIN**

Diadema antillarom

Thankfully, this species is making a comeback after an epidemic wiped out almost 90 per cent of the population in the early 1980s. Typically covered in long black or silver spines, it is found in all habitats and at all depths, throughout the entire range. Wounds from the spines may turn septic.

1

2

3

4

5

6

7

8

1 REEF URCHIN

Echinometra viridis

A relatively small, short-spined sea urchin growing to around 5cm (2in), this species has a dark reddish-brown or purple shell and paler spines, sometimes with violet tips. Fairly common in the Caribbean, it hides during the day and forages amongst the algae at night.

2 COMMON ARBACIA URCHIN

Arbacia punctulate

Growing to around 10cm (4in), this species is reddish-brown or purple with different-coloured sharp spines. It is fairly widespread throughout the region and can be found in most habitats, feeding on algae and hiding during the day.

3 VARIEGATED URCHIN

Lytechinus variegatus

This common sea urchin is covered in short sharp spines and has variegated colours. It often has bits of algae, seagrass parts, shells and small bits of coral rubble on its body, which it uses to shelter from direct sunlight. It occurs in most habitats, but prefers algae and seagrass beds. It is fairly common everywhere.

3

4 WEST INDIAN SEA EGG

Tripneustes ventricosus

A large round sea urchin, growing over 10cm (4in) in diameter, this species is dark purplish-black with short white spines. It has a symbiotic relationship with the Squat Urchin Shrimp (see page 156), which hides amidst its tentacles and spines and is very difficult to find. It is found in all of the region, as far south as Brazil and north to Bermuda.

5 CARIBBEAN SLATE PENCIL URCHIN

Eucidaris tribuloides

A very distinctive small sea urchin, this species grows to around 5cm (2in) including the short blunt spines, which are often covered in detritus. It occurs in seagrass and algae beds and coral rubble areas. Quite shy, it only ventures out later in the day.

6 BURROWING RED HEART SAND URCHIN

Meoma ventricosa

This wonderful urchin grows to around 15cm (6in) in length. It is dark purplish-red, covered in short hair-like spines, with a pentagonal pattern on the top of the test (hard shell). Quite common over the entire range, it lives under fine sand and seagrass beds where it forages, and hides away during the day. Often its broken remains are all that can be found.

7 SAND DOLLAR

Clypeaster subdepressus

A flatter cousin of the heart urchins, this species can grow to 10cm (4in) in diameter and has the same obvious pentagonal design on the top of the test. It is brownish in colour.

8 SIX-KEYHOLE SAND DOLLAR

Leodia sexiesperforata

This species has six 'keyholes' that go through the shell. It is covered in very dense short fuzzy hairs and is considered quite rare (I have only ever found the test).

1

2

3

4

5

6

7

8

Sea Cucumbers

Class: Holothuridea

SEA CUCUMBERS ARE CURIOUS echinoderms that are generally sausage-shaped with a mouth at the front and anus at the rear. They have leathery skin, and in some species the five rays typical of echinoderms are evident.

1 TIGER TAIL SEA CUCUMBER

Holothuria thomasi

Growing to 1.8m (5ft 11in) long, this species can be found in most habitats. Its mouth is on its underside and there is a series of small tentacles around the mouthparts. It extends its body out to feed on fine algae-covered sand particles, keeping its posterior firmly hidden from sight in the reef. Usually only seen at night, it is uncommon in the region.

2 BEADED SEA CUCUMBER

Euapta lappa

Long, thin and distinctly floppy, this species has a sticky and very retractile body that can stretch to over 45cm (1ft 3in). It takes in sea water and its long feathery tentacles rummage over the reef, searching for food particles to feed into its mouth. It disappears into its coral recess if disturbed. It is relatively common in the region, but is only seen at night.

3 FURRY SEA CUCUMBER

Astichopus multifidus

This species is large at around 45cm (1ft 6in) long. Its brown and pale spotted skin makes it difficult to see, and numerous lumps, bumps and spiky protuberances all over the body add to its camouflage. An active species, it lives in sandy areas around healthy reefs and in lagoons.

4 FIVE TOOTHED SEA CUCUMBER

Actinopygia agassizii

This species is quite robust and can grow to around 25cm (10in). With its dark upperside and orange-red underside, it is quite easily seen, but often only part of the body is visible as it forages amongst the coral rubble and algae beds. It is found in most sandy areas.

5 CLARK'S SEA CUCUMBER

Isostichopus macroparentheses

This species is quite small at 12cm (5in). It is cream-coloured and covered in brown warty tubercles and pale brown blotches and paler spots, with pointed white papillae. It occurs in a number of areas including Florida and most of the Gulf of Mexico and Caribbean.

6 PYGMY SEA CUCUMBER or SEA ROLLER

Aslia pygmaea

This small sea cucumber only grows to around 7cm (2¾in). It is brown with five ridges of double-rowed tubercles which are paler than the body. It is known to roll over the seabed in tidal currents. It is fairly common in Florida waters and especially at Blue Heron Bridge.

7 BLACK SEA CUCUMBER

Holothuria mexicana

Sometimes referred to as the Donkey Dung Cucumber, this species can grow to 50cm (1ft 8in) long. It has variable coloration, but is usually dull brown, grey or black with a pale orange underside and numerous tube feet. It can be found in mangrove beds, seagrass and coral rubble usually between 9–18m (30–60ft), and is widespread through the entire Caribbean region, south to Brazil and east to the Azores.

1
2
3
4
5
6
7

WORMS

Taxa: Platyhelminthes, Nematoda, Annelida, Chaetognatha, Hemichordata, Phoronida and others

THERE ARE MANY DIFFERENT species of worms to be found in the Caribbean and they are as diverse as they are interesting. There are errant worms, free spirits to wander at will; the nemertid worms, which are unsegmented and have eyes and a mouth and are active hunters on the reef; polychaete worms that generally produce a parchment or calcified tube, which they can hide inside when not feeding; bristle worms that look like centipedes and are active predators on the reef; and annelid worms, which burrow in soft sand and mud.

Errant Worms

Family: Amphinomidae

1 BEARDED FIREWORM

Hermodice carunculata

The most common of all the errant worms to be found in the Caribbean, this worm grows to 15cm (6in). It is a voracious hunter and scavenger, able to scent a decaying carcass on the reef, whereupon large numbers will descend to devour it. It occurs in all areas of the reef, but must be avoided as the tiny bunches of white bristles along its flanks can easily embed in human skin and cause extreme irritation, hence the name. It is circumtropical and is quite at home on a tropical reef, in the Mediterranean or the Red Sea.

1

2 CARIBBEAN RED-TIPPED or BLACKLINE FIREWORM

Chloeia viridis

Closely related to the Bearded Fireworm, this species can grow to around 12cm (5in). It has clumps of bristles along its red and brown banded flanks; the back has a double row of whitish spots on a darker background. It is an active reef predator and scavenger.

Flatworms

Order: Polycladida

FLATWORMS ARE CARNIVOROUS animals with a leaf-like, flat body. They undulate over the seabed by the aid of tiny cilia that beat rhythmically on their underside. The mouth and anus are located on the underside and the head has two lobed projections, which form antennae. They are distinguishable from nudibranchs by their lack of gills.

3 CARIBBEAN FLATWORM

Pseudobiceros caribbensis

A small flatworm at only 2½cm (1in), this species is predominantly brown with numerous white spots, blotches and other markings. It has a pair of pseudo-tentacles that have white tips. Due to its very small size, it is very hard to find.

4 CARIBBEAN LEOPARD FLATWORM

Pseudocerus pardalis

Very distinctive with its dark maroon coloration, darker skirt and numerous spots, this species grows to 13cm (5in). It occurs on all types of coral reefs and is considered rare in the range, as it is quite shy and tends to hide away.

5 CROZIER'S, STRIPED or TIGER FLATWORM

Prostheceraeus crozieri

This flatworm can grow to 5cm (2in). It is usually silvery grey or creamy white with thin dark lines and dark spots, and a white edge to the mantle. Its pseudo-tentacles have dark tips. It can be seen in most areas of the range.

1

2

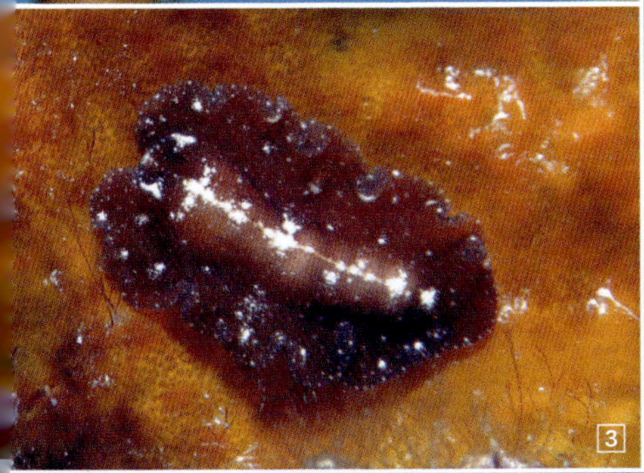
3

4

5

Feather Duster and Calcareous Tube Worms

Families: Sabellidae, Serpulidae, Terebellidae, Phoronidae

THESE WORMS LIVE within the reef and are sedentary. Most build parchment-like tubes out of very fine sand particles in the corals or on coral rubble just hidden below the sand, into which they can withdraw should danger threaten. They are light- and pressure-sensitive, so great care and patience is needed to approach them for observation or photography. Some have an operculum or hardened trapdoor, which they pull over when they hide; others just simply vanish!

1 MAGNIFICENT FEATHER DUSTER

Sabellastarte magnifica

The largest of the tube worms, this species appears as a double, spiral fan that can be 15cm (6in) across. It is found in a variety of habitats from wrecks to coral reefs, piers, pilings and coral rubble areas, usually in depths under 18m (60ft), and is fairly common in the Caribbean and Gulf of Mexico.

2 SOCIAL FEATHER DUSTER WORM

Bispira brunnea

One of the most common of all the feather duster worms, this is also the most obvious, as it grows in large social communities. Each feather cup is about 3cm (1¼in). Colour variations are common, but the feathery tentacles are typically pale pink on the outside, becoming pale purple on the inside. It is found in most habitats all over the Caribbean, but does prefer moving water.

3 STAR HORSESHOE WORM

Pomatostegus stellatus

A brilliant orange feather duster worm that forms a horseshoe shape. It has a band of white at the end of the feathery tentacles, which are tipped in orange. It is found everywhere, including among healthy stony and brain corals as well as algae and coral rubble areas. Empty tubes from dead worms become homes for a number of blennies.

4 CHRISTMAS TREE WORM

Spirobranchus giganteus

This tube worm's spiral forms the classic Christmas tree shape, growing to around 3½cm (1½in). When retracted, its operculum has a pair of spiky 'horns' that will deter any aggressors! It is multicoloured and its tubes are overgrown with live hard corals. Often found in large numbers together, this circumtropical species likes well-aerated waters, but is pressure-sensitive and will quickly retreat into its tube if you get too close.

5 YELLOW FANWORM

Notaulax occidentalis

A small yellow fan worm growing to around 3cm (1¼in), this species often lives in small groups together. It has a couple of dark bars around the delicate fan and is very shy and sensitive to light and pressure changes. It prefers coral rubble areas and is fairly common throughout the entire region.

6 SPLIT CROWN FEATHER DUSTER

Anamobaea orstedii

This feather duster worm has an obvious 'split crown' with both halves folding into the centre. It comes in a wide range of colours from translucent to vivid red, brown and tans with lighter ribs midway along each feathery radiole. This is a common species found in most habitats, and is quite shy if you get too close.

7 BROWN FANWORM

Notaulax nudicollis

Very similar to the Yellow Fanworm, this species is translucent brown with pale cream tips to the fan. It has a similar lifestyle and habitat and range to its cousin.

8 SHY FEATHER DUSTER

Acromegalomma sp.

Very similar to many others in this group, this feather duster worm has a distinct V-shaped fold. Its tube is quite distinct. The fan can be 2½cm (1in) across and is usually uniform dark purple. It is fairly common through the region, but is often mistaken for the Variegated Feather Duster.

1

2

3

4

5

6

7

8

1 **VARIEGATED FEATHER DUSTER**

Bispira variegata

Fairly common in the region, this is a small, shy feather duster, only reaching around 3cm (1¼in). The radioles are much finer than in other species. Colours are usually uniform, but it can be banded in lighter and darker shades and the eyespots may be visible in some cases. It is usually found in coral rubble areas.

2 **CALIFORNIA HORSESHOE WORM**

Phoronopsis californica

This is an invasive species and quite unlike any of the usual fan worms found in the Caribbean region. It can reach sizes of over 30cm (1ft) and has around 1,500 tentacles which form an extended spiral as a twin column. More at home in California when it was first described in 1930, this species is found on the eastern Atlantic and into the Mediterranean. This specimen was photographed at Blue Heron Bridge on the east coast of Florida.

3 **CARIBBEAN SPAGHETTI WORM**

Eupolymnia crassicornis

With stringy white tentacles that can stretch over 45cm (1ft 6in), the main body of this worm is hidden under coral rubble. It extends its sticky tentacle to collect food particles, which it brings to its mouthparts. Fairly common in the entire region, it has cousins in all of the Earth's oceans.

4 **SEA FROST**

Salmacina huxleyi

This species lives as a series of tiny worms, which form calcareous 15cm (6in) tubes that become intertwined with each other and eventually form a very fragile clump of white tubes with tiny reddish fan tufts at the ends. The tiny crowns of radioles are almost invisible. Common in all of the region, this species also occurs worldwide.

MOLLUSCS

Phylum: Mollusca

THE MOLLUSCS FORM the largest and most diverse group of marine animals on the planet and probably the most familiar group of invertebrates. The phylum includes chitons, nudibranchs or sea slugs, snails, bivalves, octopuses and squid.

Snails or Gastropods

Class: Gastropoda

THERE ARE SOME 75,000 marine species of these single-shelled molluscs worldwide, representing eight different families. Characteristics common to most species include an external shell, a soft body with a muscular foot, a rasping, tongue-like radula for feeding, and gills of some kind for respiration.

5 **WEST INDIAN FIGHTING CONCH**

Strombus pugilis

This is a small conch at around 10cm (4in) with 8–9 distinct, bluntly spiked whorls and a wide aperture. The skin is covered in fine dark dots. The operculum is sickle-shaped, as with all of the conch species. It is rare in most of the Caribbean range. Living in shallow water, local populations can be decimated by hurricanes.

6 **QUEEN CONCH**

Aliger gigas

Conches are synonymous with the Caribbean and many species play an active role in the human food chain, none more so than the Queen Conch, which is considered a delicacy. This is a large shell growing to over 21cm (9in). If you look closely, you will see its long eye stalks. Although it is fairly common in Florida and the Caribbean, there is significant concern about its population due to overfishing.

7 **ROOSTERTAIL CONCH**

Aliger gallus

Another smaller conch that grows to around 10cm (4in), its shell has a long extension at either end of the aperture's lip. It is considered fairly rare in the Caribbean due to overfishing, and tends to be found in seagrass beds and shallow patch reefs.

1

2

3

4

5

6

7

1 WEST INDIAN CHANK

Turbinella angulata

Reaching well over 30cm (1ft), this is a very large and distinctively shaped shell with 8–10 prominent and angled rib whorls, although these may be obscured by debris. It has a long tapering aperture. It can be found in most depths and inhabits a wide range of habitats, including mangrove lagoons and seagrass beds.

2 CARIBBEAN FLAME HELMET SHELL

Cassis flamea

Growing to around 15cm (6in), this species has a creamy shell, covered in wavy brownish red markings. The aperture is flattened, but does not extend as much as that of the King Helmet. It is a sand seabed dweller, often at the edge of coral reefs, and it will hide under the sand during the day and forage at night.

3 KING HELMET

Cassis tuberosa

The adult of this helmet shell reaches almost 30cm (1ft). It has a wide and flattened aperture and has dark spots or stripes around the rounded lip. The cream shell is covered in wavy brown markings. It emerges at night from soft sand to forage. This shell has been used for generations to be carved as a cameo for jewellery and decorative lamps (below).

4 RETICULATE COWRIE-HELMET

Cypraecassis testiculus

This mollusc grows to around 10cm (4in) but is usually only half this size. It has a clean shell, orange-tan in colour with longitudinal ridges all over the rounded shell. Other darker markings also traverse the shell and the pronounced rounded lip is dotted with five or six pairs of dark spots.

5 ATLANTIC TRITON'S TRUMPET

Charonia variegata

One of the largest shells to be found throughput the entire region, this species is threatened due to over-collecting for the curio trade. It has an elongated sharp spire with a number of whorls and is coloured in wavy lines of brown and cream, although older specimens may be covered in crustose algae. Like its tropical cousins, this species eats starfish and sea cucumbers at night.

6 TWIN-CORDED LATIRUS

Hemipolygona carinifera

This species has a creamy yellow shell that is usually overgrown with crustose algae. Should you be lucky enough to see the animal within, it is brilliant red. Found on most types of coral reef structures, it is quite shy and hides in the corals during the day, only venturing out at night.

7 TRUE TULIP

Fascioilaria tulipa

This shiny shell can grow to around 15cm (6in). It has mottled brown and cream markings, turning more red at the end of its spire. The front of the aperture is slightly upturned. It is considered relatively rare in the Caribbean, although I found it quite common locally at Blue Heron Bridge in eastern Florida.

8 GIANT TUN

Tonna galea

This species can grow as large as 30cm (1ft), but has no protective operculum, making it easy prey for octopuses, which can sometimes be found hiding in the empty shell. It spends much of its time buried in fine sand or mud and comes out at night, when it is an active feeder on sea cucumbers.

1

2

3

4

5

6

7

8

1 WHITE-SPOT MARGINELLA

Marginella gattata

Only growing to around 2.5cm (1in), this small mollusc has a yellow-orange shell, covered in white spots, with a wide cream skirt. I have found it to be relatively common on night dives in most areas of the Caribbean zone.

2 NETTED OLIVE

Oliva sayana

This shell can grow to around 5cm (2in) and is cylindrical with about four shallow whorls and a blunt spire. It is creamy pink and covered in a fine net-like pattern. The animal within is pinkish-mauve with fine pale markings. It is fairly common in the entire region.

3 CARIBBEAN MEASLED COWRIE

Cyprae zebra

The shell of this large cowrie is covered in white spots and indistinct darker bands on a tan or brown background. It is always shiny, because the greyish mantle extends completely over it.

4 ATLANTIC GREY COWRIE

Luria cinerea

This cowrie has a banded brownish shell, often covered in dark flecks. The mantle can completely cover the lustrous shell and is covered in small projections making it quite indistinct, or almost like a small anemone. It lives amidst the corals and is usually hidden well away during daylight hours. It is fairly common in the area.

5 ATLANTIC YELLOW COWRIE

Naria acicularis

At around 2.5cm (1in) long, this small cowrie has a yellow to pale brown shell covered in lighter spots, and can extend its brownish mantle completely over the shell. The mantle is covered with small, spiked projections. Uncommon, it prefers a coral rubble area where it hides under stones.

6 COLOURFUL MOON SNAIL

Naticarius canrena

This round shell grows to around 7cm (2¾in) and has thin bands of brown and cream, often with wavy dark lines between the bands. The mantle is pale and may have either brown or red markings and spots all over. It spends most of its time hidden under fine sand and, although it is fairly common, this night-time predator is rarely seen. It lays eggs in distinctive casings known as 'sand collars' (below).

7 WEST INDIAN TOP SNAIL

Cittarium pica

A common shoreline mollusc, this snail can be quite robust and can grow to around 13cm (5in). The shallow conical shell is black and white. Collected by many Caribbean island locals as a source of food, the animal is often referred to locally as a 'whelk'.

8 MOTTLED TOPSHELL

Calliostoma jujubinum

This shell is orange-red and can grow to around 3cm (1¼in) high. Its dark mantle is covered in small pale spots. Most often found at night when it roams amidst the coral rubble and algae beds, it is fairly rare in the region.

1
2
3
4
5
6
7
8

1 FINGERPRINT CYPHOMA

Cyphoma signatum

Slightly rarer than *C. gibbosum*, the mantle's markings resemble a human fingerprint, hence the name. The shell is a pale cream, with the mantle a lovely gold with intricate black lines. It grows to around 2.5cm (1in) and feeds on gorgonian sea fans and rods.

2 FLAMINGO TONGUE

Cyphoma gibbosum

As with cowries, this species has a lustrous shell as the mantle is able to completely cover it. The ochre shell has a high mid-ridge and a longitudinal paler stripe. The mantle is a distinctive yellow with darker yellow irregular spots lined in black. It is a popular subject to photograph as it is always found on various sea fans and rods.

3 SINGLE-TOOTH SIMNIA

Simnia uniplacata

This is a small shell, only growing to around 2cm (¾in), and takes on the colour of its host, which is usually a purple sea rod. The unmarked mantle is the same colour as the shell. Rather rarely seen due to its size and camouflage, it is local to the central Caribbean from the Bahamas to Honduras.

4 BAHAMA SIMNIA

Cymbovula acicularis

This simnia species is 2cm (¾in), and has a pale glossy shell and pale mantle with a thin orange or pink line running along the edge.

5 PINK CIRCLED SIMNIA

Cymbovula sp.

Once thought to be a juvenile Flamingo Tongue, this simnia's spots are pale pink, ringed in darker pink or brown. The elongated shape and small size of only 2cm (¾in) are distinctive. It can be found on a variety of sea whips and rods in most areas of the central Caribbean.

6 EMERALD NERITE

Smaragdia viridis

This small round shell grows to around 1cm (⅓in) and is an unmistakable vibrant brilliant green. The shell also has pale wavy markings near the spire. The eyes are black. Whilst fairly common in the Caribbean zone on seagrasses and various algae, it is rarely seen due to its small size.

7 HARLEQUIN MITER

Vexillum histrio

This mollusc can reach 2cm (¾in) long. Bold black and white patterns transverse the ribbed, red shell; the black mantle has ringed tentacles. It is fairly common in a number of areas and in a number of different habitats.

8 WIDE-MOUTHED PURPURA

Purpura patula

Usually around 5cm (2in) long, the outer shell is always a very plain lime concretion, but the underside of the shell has a large, wide aperture, the inside of which is brightly coloured in browns and blacks. It enjoys very shallow water in the intertidal zone, where it clings, limpet-like, to the ironshore rock. When threatened, it is able to exude a purplish dye.

9 BLEEDING TOOTH

Nerita peloronta

This little winkle has one or two very obvious 'bleeding teeth' at the edge of its operculum. Found from Bermuda to the West Indies, this lovely little shell rarely gets over 25mm (1in) round.

1
2
3
4
5
6
7
8
9

1 **LONG-SPINED STARSHELL**

Lithopoma phoebium

This distinctively shaped shell grows to around 5cm (2in) in diameter and has triangular, rounded spines at the outer edges of its whorls. It is yellowish brown, but always covered in detritus. It occurs from Florida and into the eastern and central Caribbean.

2 **STOCKY CERITH**

Cerithium litteratum

This mollusc can grow to 3cm (1¼in). The highly ribbed shell is usually black and white. It is common in all areas and will congregate to feed on carrion.

3 **ANTILLEAN CONE**

Conus dominicanus

This cone shell can reach 5cm (2in) long. It has a distinctive pattern of brownish-red and white in irregular patches. The name indicates its origins, but it is quite rare and was highly prized by collectors several centuries ago. It occurs in the lower and eastern Caribbean.

4 **ANTILLEAN FLESHY LIMPET**

Lucapinella limatula

This is a 'keyhole' limpet fairly common in the eastern Caribbean around the Florida coast. It has a small shell at only around 1cm (⅓in), with a hole on the top that helps direct waste water away from the snail's head and gills. It is able to extend its fleshy mantle completely over the shell, hence its common name.

Bivalves

Families: Limidae, Spondylidae, Pteriidae, Pinnidae, Arcinae

BIVALVES ARE MOLLUSCS that have two usually equal-sized hinged shells, with the animal's body hidden within.

5 **ROUGH FILE CLAM**

Lima scabra

The most colourful of the file clams, this species has a brilliant red mantle and tentacles. It is loosely attached and can 'swim' in jerky movements by opening and closing the equal-sized valves rapidly – but it soon runs out of steam!

6 **SPINY FILE CLAM**

Lima lima

As the name suggests, this clam has sharp little spikes over its roughly ribbed shell. Usually the shell is open, showing the lavender or purplish mantle. Common in most areas, the shell is often completely covered in detritus.

7 **ANTILLEAN FILE CLAM**

Lima pellucida

A small bivalve only growing to around 2.5cm (1in) long. It has a rounded white or cream shell with equal-sized halves, and has protruding white tentacles. It can 'swim' away from danger with small jerky movements by opening and closing the valves repeatedly. It is rarely seen due to its small size and nocturnal habits.

8 **ATLANTIC WINGED OYSTER**

Pteria colymbus

A small bivalve at only 7.5cm (3in), this osyter has a distinctive 'wing' shape. It is always dark-coloured, ranging from brown to black. Usually found on sea whips and plumes, it is often covered with bits of algae or even sponges, making it difficult to see; however, it is common through the entire region.

1

2

3

4

5

6

7

8

1 AMBER PEN SHELL

Pinna carnea

This pen shell can grow to around 15cm (6in). It is translucent grey with short projections all over the ribbed shell. It is usually found in sand or coral rubble areas, resting with its base well embedded out of view, and is fairly common through the entire region.

2 ATLANTIC THORNY OYSTER

Spondylus americanus

This oyster's mantle varies in colour, and its shell is often completely covered in hydroids, bits of sponge, algae and even barnacles. It is therefore rarely seen unless the valves are open for feeding; it will close if approached. It is common in the region, although often overlooked due to its camouflage.

Chitons

Class: Polyplacophora

3 ORANGE BRISTLE CHITON

Acanthochitona hemphilli

Growing to around 2½cm (1in), this is a bristly chiton, characterised by the short bristles that grow in tufts on each side of the body segment plates. It is orange-brown with reddish heart-shaped spots along the back. Usually found in fairly shallow water where there is lots of water movement, it is most frequent in the central Caribbean.

Cephalopods

Class: Cephalopoda

THE CEPHALOPODS ARE extremely intelligent and adaptable molluscs. They are free-swimming, fast-moving predators with well-developed eyes and sensory systems. All cephalopods are able to change their shape and coloration to suit their environment or to display to a mate, rival or would-be predator. Octopuses are common throughout the Caribbean zone and western Atlantic.

4 CARIBBEAN REEF OCTOPUS

Octopus briareus

This is the most common of the octopus species throughout the Caribbean range. Its camouflage techniques are legendary, and it can exhibit colour variations from a brilliant blue to multicoloured flecks of browns and creams. It has a dark ring around its eye and a very wide membrane between its eight legs, which it spreads to envelop prey. It is generally only active at night.

5 COMMON OCTOPUS

Octopus vulgaris

This octopus has no dark line around the eye and its tentacles have double rows of suckers. It enjoys seagrass beds and shallower coral reefs which adjoin a sandy or coral rubble plain. Most common in the eastern Caribbean, Florida and the Bahamas, it is also common in the eastern Atlantic and Mediterranean.

6 WHITE SPOTTED OCTOPUS

Octopus macropus

This species can reach a total length of 60cm (2ft), with tentacles four or five times the length of the body. It is orange-red, and completely covered in numerous white spots. Found on lower reefs near seagrass beds, it is present on both sides of the Atlantic and in the Mediterranean.

1
2
3
4
5
6

1 BROWNSTRIPE OCTOPUS

Octopus burryi

A small species, growing to around 30cm (1ft) in length, this species is mainly brown with obvious white markings on the body, in front of the widely spaced eyes. The skin can change structure, exhibiting wart-like projections. It prefers seagrass beds and, unlike other species, the female cares for her small eggs by carrying them in her arms.

2 CARIBBEAN TWIN-SPOT OCTOPUS

Octopus hummelincki

This species is only 25cm (10in) long, and has a knobbly body with reticulate markings. It has an additional false eye marking below the actual eye. It is found through the Caribbean and Gulf of Mexico; I have encountered it around Florida and the Virgin Islands.

3 CARIBBEAN REEF SQUID

Sepioteuthis sepioidea

A common squid, this species grows to around 30cm (1ft). It has a cylindrical body, edged in a long fin that tapers to a point at its rear. Highly variable in colour, it can often be found in small groups near the surface over coral reefs and in the shallows. It is attracted by divers' lights at night and will 'perform' to these lights and appear quite unafraid. It is common in all areas.

4 GRASS SQUID

Pickfordiateuthis pulchella

This very small squid, around 2.5cm (1in), has obvious widely spaced chromatophores dotted all over the body. It is found amongst seagrasses and shallow patch reefs which adjoin the sandy plane, but is rarely seen due to its small size and nocturnal habits.

Seahares

Order: Aplysiida

HERBIVOROUS SEAHARES are characterised by rather fleshy bodies, which are swollen towards the rear, and a head that is split into pairs of lobed oral and sensory tentacles. Sea hares can release clouds of purplish ink as well as milky white, sticky threads as a means of defence.

5 STRIATED SEAHARE

Stylocheilus polyomma

This seahare grows to about 5cm (2in) long. The body is covered in many fine brown stripes. It has a pair of long tentacles and a pair of dark brown eyes to the front of these. It can often be found in large breeding aggregations when many individuals collect together on various type of algae on the coral rubble seabed, or amongst algae and seagrass meadows.

6 SPOTTED SEAHARE

Aplysia dactylomeia

At around 20cm (8in), this is quite a large and distinctive seahare, with a variegated body that is covered in dark circles or spots. The mantle 'wing' flaps are usually folded up and over the animal, but it does have the ability to 'fly' by undulating its wings.

7 BABA'S SEAHARE

Petalifera ramosa

This is a stunning, almost translucent green seahare that can grow to at least 7cm (2¾in). The body is covered in an intricate pattern of pale knobs, surrounded by a pattern of fine white lines. It has two long oral tentacles and its small eyes are directly to the sides of the main tentacles on the head. It is found in shallow waters around seagrass and algae beds and is uncommon.

8 WHITE-SPOTTED SEAHARE

Aplysia parvula

This small seahare grows to around 4cm (1½in) long and is mottled cream or brown, with an obvious line behind its tentacles that curves around the front of the body. The gills are situated high along its back, reminiscent of a nudibranch. It is found from the Cayman Islands to Honduras.

1

2

3

4

5

6

7

8

Nudibranchs

Order: Nudibranchia

THE NAME 'NUDIBRANCH' means 'naked gills'. These are shell-less gastropods, and vary enormously in shape and colour. However, they usually have obvious rhinophores at the head and a set of gills at the rear of the body, often various projections on their bodies, and all are carnivorous. Many species have undergone several changes of scientific name, as DNA study has clarified their relationships, and further name changes will be updated in future issues of this book.

1 GOLD-LINED SEA GODDESS

Felimare ruthae

Growing to around 2.5cm (1in), this is a beautifully coloured nudibranch. It is considered fairly rare in the Caribbean, but can be locally common when the blue sponges that they feed on are abundant.

2 ZEBRA DORIS

Felimare zebra

This species reaches over 2.5cm (1in) and has an orange body streaked with blue lines and wavy markings. The outer margin turns to pale blue-grey with black spots, changing to an orange outer margin. It is widespread along the western Atlantic, including Bermuda.

3 LEATHER-BACKED DORIS

Playdoris anguistipes

A dark reddish-brown nudibranch, this species is covered in small white specks and has two dark rhinophores. The gills are a brownish fringe. It feeds fairly exclusively on red sponges and is widely distributed throughout the entire region.

4 BLACK-SPOTTED NUDIBRANCH

Ceratophyllidia papilligera

Spread through the entire Atlantic, this species is white with numerous rounded cerata and irregular black spots. It occurs along the west coast of Africa, as well as Florida, through the Cayman Islands and to Honduras.

5 RED CORAL NUDIBRANCH

Plocamopherus lucayensis

This species is vibrant red with a ring of white-tipped cerata at the front, and also white tips to the gills, which are set quite far back on the body. I found this in the Cayman Islands, but they are recorded elsewhere feeding on sea rods.

6 EYESPOT COSTASIELLA

Costasiella ocellifera

Often referred to as the 'Shaun the Sheep' nudibranch, this tiny 1.3cm (½in) slug has numerous large inflated cerata along the back which go through many colour changes before becoming white with dark brown tips. The head has a prominent iridescent blue spot and there are other blue markings on the body.

7 WHITE-SPECKLED AEOLID

Paleo jubatus

This species can grow to around 5cm (2in). Generally cream in colour, it has numerous rows of cerata down either side of the back and a white line and spots down the centre. Common in most areas, it is found almost exclusively on knobby sea rods (see page 138), feeding on the polyps, but I have also found it on a variety of sea fans too. It may be found at all times of day, but is certainly more obvious at night.

8 CRISSCROSS TRITONIA

Tritonia bayeri

Only growing to around 2cm (¾in), this species is translucent white with widespread spiky cerata and a curious network pattern on the back. It is widespread and feeds on gorgonians, hydroids, sea whips and the like.

1
2
3
4
5
6
7
8

1 LONG-HORN NUDIBRANCH

Austraeolis catina

This nudibranch can reach 4cm (1½in) in length. It has six pairs of clumped cerata, which have white speckles and are often two-tone white and cream, down its creamy tan-coloured back; the front tentacles are very long and striped. It is fairly common in the region and can be found on hydroids and coral polyps on healthy, living reefs.

2 FROSTY MORDILLA

Noumeaella kristenseni

Only 1.3cm (½in) long, this pale cream sea slug is covered in very fine white spots and has light brown markings down its sides, below the groups of eight or nine cerata. Reasonably common in the region, its size enables it to hide very effectively amongst the hydroids that it feeds on.

3 PAPILLATE FACELINA

Palisa papillata

This nudibranch grows to over 1.5cm (½in). It has a creamy grey body with darker markings and the groups of cerata are blueish at their base. Numerous white spots and splotches are also present. It feeds on a variety of hydroids. It is fairly rare in the region, but more common in the central Caribbean, and was first described from Jamaica in 1964.

4 DUSHIA FLABELLINA

Flabellina dushia

This sea slug grows to around 2.5cm (1in). It has a white body and clusters of translucent cerata which are striped with orange, red and white. It has a long tail and feeds on hydroids. It is reasonably common in the central Caribbean.

5 LONG-EARED FAVORINUS

Favorinus auritulus

A small nudibranch at only 1.3cm (½in), this species is brown and white with translucent brown and white clusters of cerata and long ridged, smooth rhinophores. It may have pale blue markings at the front of the head and at the tip of its long tail. It is seldom seen in the region, possibly due to its small size, but is more common in the central Caribbean.

6 BERGH'S AEOLLIDIA

Berghia benteva

This species grows to around 2cm (¾in). It is orange-tan with a broad white stripe along its back. The rhinophores are long and ribbed and the body's sides are covered in numerous long cerata. Also found on the coast of West Africa, it occurs in numerous habitats.

7 TASSELLED NUDIBRANCH

Bornella calcarata

Growing to about 4cm (1½in), this mainly brown nudibranch has a net-like pattern over the body. The highly branched cerata form small rows along the body. It enjoys a number of habitats and its shape and colour help to camouflage it as it feeds on various hydroids. It is found in a number of localities in the region.

8 SPOTTED THECACERA

Thecacera pennigera

This species has a very distinctive coloration (and shape), being white with widely spaced orange and black spots all over the body. Its rhinophores are in obvious sheaths. Found on both sides of the Atlantic, it also occurs in the English Channel, Mediterranean and the Bahamas.

1
2
3
4
5
6
7
8

1 WARTY SIDEGILL SLUG

Pleurobranchus areolatus

This large species grows to around 10cm (4in). It is quite rounded with numerous warty tubercles and is usually mottled brown and cream. It is regarded as uncommon, but is also recorded from the eastern Pacific. I found it at night in an inner lagoon amongst the seagrass beds and sheltered sandy areas on Cayman Brac.

2 APRICOT SIDEGILL SLUG

Berthellina quadridens

A very distinctive smooth, orange slug, this species can reach 5cm (2in). It is slightly translucent, and can secrete sulphuric acid in defence. It tends to hide under corals during the day.

3 FLAPPING DINGBAT

Gastropteron chacmol

Who thinks of these common names?! This is a tiny slug, only 7mm (¼in) long. Reddish brown with a white trim to its overlapping parapodia flaps and small bluish white markings, it has a single syphon at the top of its head. It can 'fly' if danger threatens. Found on sand in varied depths, it is fairly common, but rarely seen due to its small size.

4 NORMAN'S AGLAJA

Chelidonura normani

At only 1cm (⅓in) long, this tiny critter is fairly easy to spot on a white sandy seabed, despite its small size. It may be black with small white dots, to grey-brown with light grey spots and a blue margin to the head and tail. It is fairly common in all central regions.

5 LEECH HEADSHIELD SLUG

Chelidonura hirundinina

This species can grow to around 2.5cm (1in). Brightly coloured with blue, gold and black stripes along the body, it also has a white bar near its forked tail. It is commonly found in the Caribbean zone.

6 BLACK AGLAJA

Nakamigawaia felis

This slug is 1cm (⅓in) long and all black, but much more slender than Norman's Aglaja. The two lobes over the body are of different sizes. It is found on plain sandy seabeds at a variety of depths.

7 LETTUCE SEA SLUG

Elysia crispata

This is probably the most popular nudibranch to be found in the Caribbean – loved by everyone! It can reach 5cm (2in) long and is locally common in all habitats, in most regions of the Caribbean. The enlarged and convoluted sides that extend around its head, resembling a crisp lettuce leaf, are known as parapodia and come in an amazing variety of colours. This species feeds exclusively on various green algae, but not all of the algae is digested. Lettuce Sea Slugs are 'sacoglossans', meaning they are able to store chloroplasts from algae in their parapodia and benefit from the photosynthetic energy.

1
2
3
4
5
6
7

1 STRIPED or LINED ELYSIA

Elysia prafensis

Around 2.5cm (1in) long, this slug has longitudinal stripes all along its body and the outer edge of the body ruffles have a black edge and slightly raised white bumps. It also has yellow-tipped rhinophores. It lives on a variety of algae and seagrasses, and is found in most places in the central Caribbean.

2 BROWN-TIPPED ELYSIA

Elysia scops

A small, fuzzy-looking sea slug, this species is only 1.3cm (½in) long. Its large, thick lobes are usually upright rather than folded over like other species. Its rhinophores are tipped reddish brown and covered in numerous small white lumps and bumps. It can be found on various algae, usually in depths less than 15m (50ft), in most areas of the Caribbean region.

3 SEAGRASS ELYSIA

Elysia tuca

This slug is only about 8mm (½in). It is green with a large rounded head and fairly long rhinophores, and its folded-over lobes are almost always joined together in a long wavy line, edged in cream. It is only found on Turtle Grass (see page 208) and therefore in fairly shallow water, but is often overlooked.

4 ORNATE ELYSIA

Elysia ornata

Ornate is correct! This vibrantly coloured species can grow to 5cm (2in). It is mainly translucent lime green with a high arched back. The body sides and extended lobes are covered in small black and white dots. The edges of the lobes and rhinophores are golden and trimmed in black.

5 ZEMI'S ELYSIA

Elysia zemi

Growing to just over 1.3cm (½in) long, this rusty brown sea slug has convoluted lobes, which are usually open and covered in small spiky protuberances. The rhinophores are long and there is a purplish patch on the head. It is locally common on algae.

6 ALGAE ELYSIA

Elysia patina

A distinctive species, this slug has a convoluted outer edge to the lobes, which are edged in a wide cream band. The body is mainly green but splashed with small cream or white markings. The rhinophores are long and mainly brown. It is found on numerous species of algae.

7 ARIANE'S MELIBE

Melibe arianeae

This slug was first found at Blue Heron Bridge in east Florida by Ariane Dimitris and was formally described in 2013. The photograph in this book was taken in the Cayman Islands, revealing that the species' distribution is wider than was previously thought. Being almost transparent and quite small, it is no surprise that this peculiar little sea slug has been overlooked before now.

8 BLACK-SPECKLED DORIOPSILLA

***Doriopsilla* spp.**

This nudibranch's identity is often confused, as there are several species which all look very similar. Lacking a radula (rasping mouthparts), they feed by dissolving their sponge food source externally with the aid of special enzymes and then ingesting the sponge material, minus the sponge's skeleton. Regarded as uncommon in the region, the Black-Speckled Doriopsilla may be more prolific locally, when its sponge hosts are more available.

1
2
3
4
5
6
7
8

TUNICATES

Subphylum: Tunicata

ALSO REFERRED TO AS ascidians or sea squirts, tunicates are very closely related to vertebrates, as they have a notochord (the precursor to a spinal cord), which is more obvious in the planktonic stage of the young, and their gullet is perforated like the gills of a fish.

1 GIANT TUNICATE

Polycarpa spongiabilis

Reaching 10cm (4in) in length, this tunicate has two large syphons. It is usually covered in all manner of detritus and algae, which makes for excellent camouflage. It is quite shy and will close its syphons if you get too close, rendering it almost invisible. It can be found all over the region in a large number of habitats and depths.

2 PAINTED TUNICATE

Clavelina picta

This species forms small attached colonies with each bell-shaped, translucent animal being 2cm (¾in) long, and shaded in purple or blue. It is common almost everywhere and can be found in all depths and habitats.

3 BLUE BELL TUNICATE

Clavelina puerto-secensis

Measuring 1.4cm (½in), this tunicate is uniformly bluish purple with its twin syphons edged in white. It is the smallest of the species to be found in the Caribbean and forms small clumps of individuals amongst various corals.

4 MANGROVE TUNICATE

Ecteinascidia turbinata

Very similar to the other 'bell' tunicates, this species can grow to 2.5cm (1in). It is translucent orange with both siphons at the top. It can be found mixed in with other tunicates. Colonies live amidst corals and overgrowing algae and seagrasses, and may hold hundreds of animals. It is common through the region, and not always around mangroves.

5 BUTTON TUNICATE

Distaplia corolla

Growing in small clusters which may join together, this tiny tunicate is only around 5mm (¼in) and comes in a variety of colours, but mainly purple or orange. It has a single large inhalant syphon, surrounded by several small exhalant syphons. Regularly seen through the entire region, it overgrows sponges and corals.

6 WHITE SPECK TUNICATE

Didemnum conchyliatum

This tiny yellow-white sea squirt is only 5mm (¼in). It will form small aggregating colonies, joined together or as individuals. It likes light and is usually in depths shallower than 30m (100ft) in more open locations with a slight current. It occurs in a number of habitats and overgrows sponges.

7 CARIBBEAN BLACK SEA SQUIRT

Phallusia nigra

A large sea squirt, growing to 10cm (4in), this species is black with no obvious markings. The inhalant tube is at the top and the exhalant tube branches off the side. It is found in most areas.

8 GEOMETRIC ENCRUSTING TUNICATE

Botryllus sp.

This is a small tunicate with individual animals only around ¾cm (¼in) or less. It is white with a black geometric rim and centre. Fairly common through the entire region, it is found encrusting dead coral and various sponges.

1

2

3

4

5

6

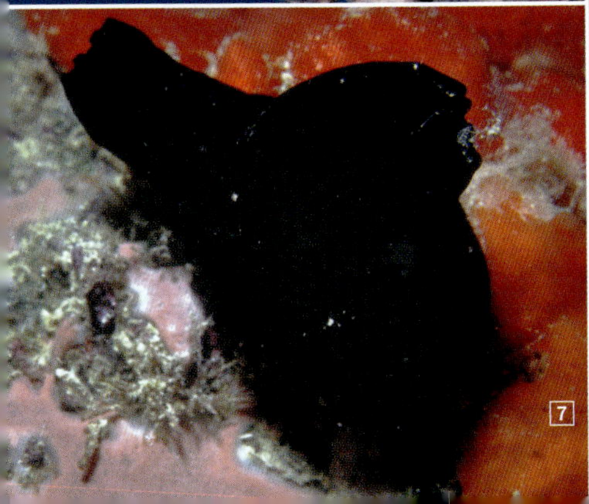

7

8

SPONGES

Phylum: Porifera

SPONGES ARE VERY SIMPLE multicellular animals. They are covered in small pores called ostia, through which they suck in nutrient-rich seawater. The waste or 'cleaned' water is expelled through larger holes called oscula. Coming in many colours and shapes, they grow to suit their environment. For example, in areas with very strong currents, large barrel sponges grow in a more flattened shape, whereas on a sheltered deep reef, they may attain their full barrel-like shape.

1 STRAWBERRY VASE SPONGE

Mycale laxissima

Usually only 30cm (1ft) long, this sponge comes in vivid reds of various shades. It has a very wide inhalant tube and a very rough bobbly texture. Overgrowing sea rods and black coral branches, it likes deep vertical walls and is a common sight on Bloody Bay Wall off Little Cayman Island.

2 PINK VASE SPONGE

Niphates digitalis

Growing to around 30cm (1ft), this sponge has a very rough, bobbly surface and, although it can be found opened up, it generally forms large upright tubes. It is associated with the Sponge Zoanthid (see page 126), which can completely inhabit the outer skin of the sponge.

3 AZURE VASE SPONGE

Callyspongia plicifera

This sponge can be 45cm (1ft 6in) high and is commonly seen in pairs. The outer form is steeply convoluted with ridges, valleys and other forms and always appears very clean. Most commonly found on walls and in depths less than 21m (70ft), it is found in most areas of the Caribbean.

4 BRANCHING TUBE SPONGE

Pseudoceratina crassa

Growing as a large cluster of tubes that can each reach 45cm (1ft 6in), this species is mainly yellow-brown or tinged with purple, and is covered in many small bumps, giving a rough texture. It tends to branch low down and stretch out into the current on reef walls. It is common in all areas, but especially the central Caribbean where the water is warmer.

5 ANTILLES STOVE-PIPE SPONGE

Aplysina archeri

Forming huge tubes over 1.8m (5ft 11in) long, this sponge grows in clumps of many individuals of different sizes, but can also be found singly. It is usually pale purple with a cream interior, and the tubes are slightly larger at the top end. It likes deep walls and shipwrecks and is common in the Netherlands Antilles.

6 YELLOW TUBE SPONGE

Aplysina fistularis

Growing in a many-branched colony over 120cm (4ft), the individuals of this sponge can be colonised by other sponges growing off them. Often tiny extensions adorn the tubes and in many cases all of them may be fused together. It likes open water, reef crests and walls, and is common in all areas.

7 CONVOLUTED BARREL SPONGE

Aplysina lacunose

This is a very thick-walled sponge, usually greenish in colour, but with a very convoluted outer face. Its folds are often filled with debris and detritus, allowing other organisms to grow on them. Enjoying a position on vertical walls in most areas of the Caribbean, it can be found in association with many other sponge species.

8 BROWN TUBE SPONGE

Agelas conifera

Similar to the Yellow Tube Sponge except in colour, the individuals of this sponge can grow over 90cm (3ft) and may be fused together at the base or grow in rows. It can be found on reef crests and walls that plummet into the depths, and is common on all deep reefs.

1
2
3
4
5
6
7
8

1 BROWN BOWL SPONGE

Cribrochalina vasculum

Quite clearly bowl shaped and a deep brown colour, this sponge grows from a single thick stem and can reach 45cm (1ft 6in). This is a solid and stiff sponge with a rough textured surface. The bowl is usually filled with detritus and fish sleep within this recess at night; however, on the deeper reef walls, this sponge forms large flattish discs. It is fairly common in the region.

2 TOUCH-ME-NOT SPONGE

Neofibularia nolitangere

This can be a very large sponge, growing to over 120cm (4ft), and may be found in pairs, in larger groups or on its own. A deep red, robust sponge, it has a very wide exhalant opening at the top and a very rough texture. The spicules are highly toxic and can cause very painful rashes and numbness – some people may have severe allergic reactions and may require medical assistance. It is found in all areas.

3 ROW PORE ROPE SPONGE

Aplysina cauliformis

This sponge extends out from the vertical wall into the current and its branches extend upwards from a single arm or several. The weight of these results in a curved shape. The oscula form long rows. Mainly red, it can also be lavender, with yellowish tinges to the inhalant openings. A widespread species, it is very common off the north shore of Little Cayman Island.

4 SCATTERED PORE ROPE SPONGE

Aplysina fulva

Arms of this sponge can reach well over 1.8m (5ft 11in). It is found on deeper reefs and walls. This was photographed at Maria La Gorda in western Cuba and was a very large specimen with many drooping arms.

5 BROWN ENCRUSTING OCTOPUS SPONGE

Ectyoplasia ferox

This species can grow to 45cm (1ft 6in), but is usually much smaller than this. It forms a large tube-like structure, wider and barrel-shaped in the centre. It may have a large exhalant opening, but is characterised by having many small, raised inhalant tubes all over its surface. It is common in the entire region and its name comes from the fact that it often overgrows dead corals, which then resemble the arms of an octopus.

6 ORANGE ELEPHANT EAR SPONGE

Agelas clathrodes

This is a very large overgrowing sponge sheet, often with a convoluted outer edge. It can easily stretch over 1.8m (5ft 11in) and is common in all areas of current. It loves deep walls and will be found in close contact with various other sponges.

7 GIANT BARREL SPONGE

Xestospongia muta

Growing around 1.3cm (½in) a year, larger Giant Barrel Sponges can easily reach over 1.8m (5ft 11in) tall and be over 120 years old. The outer lip is quite fragile and easily broken off and severe hurricanes can break the main body, but the sponge soon regenerates, growing multiple smaller sponges from its base. Lobsters and large groupers can be found in these sponges at night. It is found through the entire range.

8 ERECT ROPE SPONGE

Amphimedon compressa

Similar to the Row Pore Rope Sponge, this species is erect and usually grows from a single source, branching slightly as it reaches upwards to the surface. It can be easily over 90cm (3ft) high. Very common in Cozumel, it is found in all of the central Caribbean region.

1

2

3

4

5

8

6

7

1 NETTED BARREL SPONGE

Verongula gigantea

This is a common sponge that turtles like to eat; their large bite marks reveal bright yellow coloration underneath. It may grow as large as 120cm (4ft) high. Found amidst many other types of sponges and corals, it occurs in most areas, always below 18m (60ft) and more often at the edge of a deep reef crest where there is more water movement.

2 BROWN ENCRUSTING SPONGE

Ectyoplasia ferox

This sponge has a huge variety of forms and will grow to suit its immediate habitat, sometimes overtaking a damaged coral, or extending out into the current of a steep coral wall, or deeply branching into a weirdly shaped clump. Found in mid-water over 12m (39ft), it is common everywhere.

3 RED BORING SPONGE

Cliona delitrix

This red sponge does grow over and encrust hard corals, but also bores into corals. The excurrent openings are large and raised. It is fairly common in the region, on living reefs in clear, clean and well-aerated water.

4 VARIABLE BORING SPONGE

Siphonodictyon coralliphagum

There are numerous types of boring sponge. This species grows over a damaged area of coral in large flat sheets, but also forms rounded raised oscula. It uses a type of acid to dissolve the coral, allowing it to penetrate deep into the coral skeleton below, creating huge holes that will weaken the coral.

5 ORANGE SPIKY ENCRUSTING SPONGE

Scopalina ruetzleri

Not as common as some of the other more obvious sponges, this species can grow to around 30cm (1ft) and is covered in numerous spiky protrusions, making it look rather dangerous. It encrusts more damaged coral areas and is often covered in numerous types of hydroids.

6 HIGH-VEINED ENCRUSTING SPONGE

Spirastrella hartmani

Growing over a wide area in fairly sheltered and dark areas of the reef, this sponge can reach around 90cm (1m) across. It has very obvious large raised veins, which are vaguely star-shaped and run to and from the excurrent openings.

7 FIRE SPONGE

Tedania ignis

This small reddish-brown sponge, reaching almost 30cm (1ft), can cause a severe burn-like irritation, pain and numbness that can last several days and may require medical attention. It is common in lagoons, mangrove roots and seagrass beds.

8 ORANGE ICING SPONGE

Mycale laevis

This is a common encrusting sponge with raised transparent margins to its oscula; the inhalant openings are also very obvious. It grows underneath various corals, where it can actually alter the shape of the coral growth into a more convoluted or wavy outer edge. It occurs throughout the entire region.

8

1
2
3
4
5
6
7
8

1 **STINKER SPONGE**

Ircinia felix

This grey sponge can grow to around 30cm (1ft). It has very distinctive hexagons of small white knobs joined by a raised line, and random, large excurrent openings. It is common through the entire region in depths less than 20m (66ft).

2 **CORAL ENCRUSTING SPONGE**

Cliona langae

This brown sponge will overgrow corals which may have been affected by the cyanobacteria that causes Black Band Disease (*Phormidium corallyticum*), which can decimate hard corals. The sponge bores into the coral and takes on its underlying form. It is fairly common in the region. There are another couple of species that are almost identical to this sponge, and all have the same effect.

3 **PINK AND RED ENCRUSTING SPONGE**

Spirastrella coccinea

Growing to around 90cm (3ft), this encrusting sponge prefers reefs and walls with a more shaded aspect, in underhanging corals, caverns and under ledges. It is found locally in most areas of the region.

4 **RED COLUMN SPONGE**

Pandaros acanthifolium

This is a very dark red, upright sponge, growing from a single column in the sand or coral rubble. It will branch into two or three 'fingers'. It has a rough texture and the oscula are peppered all over. Generally uncommon in the region, it can be abundant locally.

5 **RED ENCRUSTING SPONGE**

Monanchora barbadensis

Common throughout the entire region, this bright red encrusting sponge often overgrows dead corals, but does appear to prefer more shaded areas under overhangs and ledges and caverns. It can inhibit coral growth.

6 **CRYPTIC SPONGE**

Leucetta floridana

Coming in a wide variety of pastel colours, often all jumbled together, this sponge can grow as long as 30cm (1ft). It inhabits small caverns and recesses away from strong sunlight, but appears to enjoy more current and moving water to keep it supplied with nutrients. It is fairly common in the region.

7 **ROUNDED ROPE SPONGE**

Agelas sceptrum

Growing around 90cm (3ft) long, this large, knobbly, round sponge has characteristic lighter excurrent tubes. Found on deeper reefs on walls and even under overhanging ledges, it is quite common in all of the region.

BRYOZOANS

Phylum: Bryozoa

BRYOZOANS ARE TINY SESSILE (non-moving) animals that mostly occur in shallow seas. Their colonies, composed of numerous individual zooids, grow by budding (producing clones), although they also reproduce sexually and their swimming larvae will settle and establish new colonies. They are filter-feeders and help to maintain water quality in their native habitats, though they can be destructive when invasive. Bryozoan colonies feel crispy to the touch.

8 **TAN FAN BRYOZOAN**

Scrupocellaria sp.

This species forms small crusty fans, about 3cm (1¼in) across, in areas of turbulent shallow water amidst spur and groove reefs. Locally quite common, it is often overlooked, but is found through the entire region.

1
2
3
4
5
6
7
8

PLANTS AND ALGAE

SEAGRASSES

Family: Hydrocheritaceae; Cymodoceaceae

SEAGRASSES ARE TRUE flowering plants, with stems, leaves and roots with spreading rhizomes which also help to anchor the plants to the substrate. They also produce flowers, although these are rather small and indistinct.

1 TURTLE GRASS

Thalassia testudinum

This plant with its 1m (3.25ft) flattened long leaves forms dense meadows in sheltered lagoons. It produces small flowers, and seeds in a conspicuous pod. By far the most abundant marine plant in the entire region, it provides shelter for many species and a home to countless juvenile fish.

2 MIDRIB SEAGRASS

Halophila bailonis

This plant's leaves are only around 5cm (2in) long, rounded and oval in shape with a distinctive mid-rib. It has an extensive root system. The leaves may form huge meadows, often covered in detritus and encrusting algae. This too is a nursery for many small and juvenile invertebrates.

3 MANATEE GRASS

Syringodium filiforme

This is a quite erect, thin-bladed seagrass with round green leaves. It has extensive runners, which are quite obvious on the surface of the sediment. It is fairly common, often mixed in with Turtle Grass, and is common in the region. As the name suggests, it is a favourite food for the West Indian Manatee (see page 116).

4 MANGROVES

Families: Rhizophoraceae, Acanthaceae

THE MANGROVES GROWING in the Caribbean today have been scientifically proven to be descended from plants that grew during the Late Cretaceous period (143–66 million years ago). Covering more than 14,700km² across the region, mangroves are transitional seawater–land wetland ecosystems that are vital in so many ways to the health of Caribbean reefs and islands. There are a number of mangrove species in the Caribbean – pictured is the Red Mangrove *Rhizophora mangle* with its notable reddish root props.

1
2
3
4

ALGAE

Taxa: Archaeplastida and others

GREEN, BROWN AND RED ALGAE are all found in the Caribbean zone. However, the most commonly seen (though rarely identified) algae are those single-celled species that live in the coral membrane, giving the corals their colours. It is these species that are expunged during rising sea temperatures, resulting in coral bleaching. They do grow back, but only when the water temperature cools.

Brown Algae

Class: Phaeophyceae

1 SARGASSUM SEAWEED

Sargassum fluitans

This alga forms large floating rafts, supported by pea-shaped air bladders. It is light brown to cream and has long serrated blades. Common within the entire range, it is famously associated with the huge Sargasso Sea.

2 WHITE-VEIN SARGASSUM

Sargassum hystrix

Growing in clumps to around 45cm (1ft 6in), this leaflike alga has a clearly defined light grey vein in the centre of the leaf. The clumps are firmly attached to coral structures, on deeper patch reefs that have large sandy areas around. It is common from Florida, right through the Caribbean and Gulf of Mexico.

3 SARGASSUM ALGA

Sargassum acinarium

Growing to around 5cm (2in) long, this is quite a rough-textured, tough alga with olive green-brown small leaves. Like all others in its genus, it has small gas-filled floats to keep it afloat. It is fairly common in the entire region.

4 LEAFY ROLLED-BLADE ALGA

Padina boergesenii

Although it is green in colour, with iridescent bluish-green stripes all over its curved blades, this is, taxonomically speaking, a brown alga! It grows in clumps to 15cm (6in) across, prefers a rocky substrate to attach to, and is mainly in water shallower than 15m (50ft).

5 BLUE-BANDED ALGA

Taonia abbottiana

Very similar to the Leafy Rolled-blade Alga, this species has shorter, more flattened leaves with very distinctive blue bands. It can grow to 10cm (4in) and is quite common in the region, but often misidentified.

6 LEAFY FLAT-BLADE ALGA

Stipopodium zonale

This is very similar to the Leafy Rolled-Blade Alga, except that its fronds can extend over 30cm (1ft). It is quite a bushy alga and is found in most reef systems, particularly in spur and groove reefs on hard substrate, and usually in quite shallow water.

7 BLISTERED LEAF ALGA

Turbinaria turbinata

Growing to around 45cm (1ft 6in) across, this distinctive tri-cornered alga is found in shallow lagoons and inside barrier reefs. Usually home to a number of species of fish including seahorses, this is fairly widespread through the entire region.

8 WHITE SCROLL ALGA

Padina jamaicensis

Very similar in every way to its cousins in the Mediterranean and Red Sea, this parchment-like alga is grey in colour and forms convoluted curves and semicircular forms around 15cm (6in) across. I have found it pretty much everywhere I have travelled in the region, usually in very shallow water.

9 Y-BRANCHED ALGA

Dictyota sp.

These are multicoloured small algae that can form large clumps over 45cm (1ft 6in). They are rarely, if ever, brown in colour, and their blades are forked with rounded tips. They tend to grow over a hard substrate and are very common in all depths in all regions.

1
2
3
4
5
6
7
8
9

Green Algae

Taxa: Chlorophyta and others

1 SEA PEARL

Venticaria ventricosa

Sometimes referred to as 'Sailors' Eyeballs', this spherical species is one of the largest single-celled algae in the world, with individuals reaching around 5cm (2in) across. It is usually found in small groups and is covered in Lavender Crust Algae (see page 216). Common in all areas, it is found in all depths and all habitats.

2 ELONGATED SEA PEARLS

Valonia macrophysa

Much smaller than the Sea Pearl, this species is more elongated and rarely reaches 2cm (¾in). It forms large tight clusters that may be spherical, but are usually elongated, and have a silvery sheen from overgrowing algae. It is a common species in all areas.

3 GREEN CUP

Polyphysa polyphysoides

Quite a small and inconspicuous alga, this rarely grows to over 1cm (⅓in) across. It takes the form of a small, ribbed upward-aligned, bright green cup. Sometimes confused with the Mermaid's Green Wineglass Alga, it is much smaller. Rarely seen due to its small size, it is only found in shallow water, and occurs in most areas of the tropical Caribbean.

4 THREE FINGER LEAF ALGA

Halimeda incrassata

Growing from a single stalk on rough coral gravel seabeds, this alga is darker green than other species in the same region. The leaves can often be calcified and are quite upright and three-lobed or ribbed in structure, with a pronounced midrib. This and other *Halimeda* species are major contributors to calcium carbonate on the reefs.

5 LARGE-LEAF HANGING VINE

Halimeda copiosa

This alga's leaf segments are around 2cm (¾in) across and the strands can be over 60cm (2ft) long. It prefers shaded areas on the reef and underhanging cliff walls and can form huge clumps attached to the hard substrate. It is common in the entire area.

6 SMALL-LEAF HANGING VINE

Halimeda goreaui

These small three-lobed circular leaves form long chains held together by just a thin strand. Common off Cozumel and most other areas in the zone, this alga prefers underhanging coral heads where it has more protection. It is found in all depths and the strands can reach over 30cm (1ft).

7 BRISTLE BALL BRUSH

Penicillus dumetosus

This species forms a rounded, dark green ball of tightly packed bristles. It can grow to 15cm (6in), although is usually much smaller. It is usually found in shallow waters in spur and groove reefs and inner barrier reefs. Fairly common in the region, it is usually found around seagrass beds and other patches of algae.

8 GREEN JOINT-STALKED ALGA

Halimeda monile

Fairly common in the region, this is an erect green alga. The blades are somewhat cylindrical and flattened and it grows by extending a long root or stolon. It is usually found in shallow, sheltered sandy areas, such as lagoons, or sheltered behind a barrier reef.

9 WATERCRESS ALGA

Halimeda opuntia

This is a thick and tightly branched alga, very similar to the other species in this family. It is tightly packed and often found growing in vase sponges or amidst close-packed corals. It is found in all areas of the region and is considered to be a major sequesterer of carbon, providing calcium carbonate to the surrounding coral reefs.

1
2
3
4
5
6
7
8
9

1 MERMAID'S TEA CUP

Udotea cyathiformis

This is a distinctive cup-shaped alga, growing from a single thin stalk. The outer margins are darker green than the centre of the cup, which can often get filled with debris. It grows on a variety of substrates including coral rubble as well as dead coral areas, and is commonly found throughout the region.

2 MERMAID'S FAN

Udotea flabellum

Forming semicircular and fan-shaped blades of bright green, this species can form patches over 20cm (8in) across. Often there are many individual blades clumped together with other very similar species in the same family group, but with slight variations in shape. It prefers rough sandy seabeds.

3 GREEN MERMAID'S WINE GLASS

Acetabularia calyculus

This is a small alga, only growing to around 3.5cm (1½in); its cups are 8mm (¼in) in diameter. However, it grows in larger clumps of many individuals. It is found in most regions in shallow water, but always in a protected area of reef or mangrove lagoon.

4 CARIBBEAN MERMAID'S WINE GLASS

Acetabularia crenulata

This unmistakable alga consists of clumps of thin green or white stalks, topped with a thallus – a very pale green or white, rayed disc hardened by encrusted limestone. Umbrella-shaped, it grows around 8cm (3¼in) tall with the disc being around 2cm (¾in). Large aggregations are concentrated on a stone or rocky substrate. It is quite often found on the lower shore and rarely deeper than 6m (20ft).

5 FINGER CODIUM

Codium isthmocladum

Identified by the small clumps of 'fuzzy' fingers, generally from a single or double stalk, this alga branches widely at its base to form many small rounded clumps of stalks. Quite often coated in a filamentous, slimy alga, this species enjoys aerated, shallow water conditions where wave surge and high light are present. It has an almost identical cousin in the eastern Atlantic and Mediterranean.

6 GREEN NET ALGA

Microdictyon boergesenii

Often found covering large swathes of dead coral reef or coral debris in spur and groove reefs, this bright green alga is formed of a stiff mesh of long interconnected filaments that appear to be rolled over. It is found in most areas, and in most depths up to 30m (100ft).

7 NETWORK ALGAE

Microdictyon marinum

This species is not too common in the region, but, when found, there will be a lot of it! It tends to cover dead or damaged areas of coral reef and forms convoluted or crumpled sheets of a fine green network.

8 GREEN CARPET ALGA

Caulerpa taxifolia

This invasive species is incredibly virulent and adaptable to new areas. First recorded in the US Virgin Islands in 1802, it is now found in most regions of the Caribbean. It spreads from a single stalk, and an enzyme in the alga is toxic. Leaves can be 6–65cm (2½in–2ft 2in) long. In 1m² (11ft²) of seabed, an individual plant can have a stolon or extending root of over 3m (10ft) in length, with over 200 leaf fronds. *C. taxifolia* is very similar to *C. mexicana* (see page 216), which is also found throughout the region.

1
2
3
4
5
6
7
8

1 **GREEN GRAPE ALGA**

Caulerpa racemosa

Growing to around 15cm (6in), this conspicuous alga is a lively green colour, often with a bluish tint. It has long runners that bear upright stalks, covered in round grape-like clusters. It is found in depths to 18m (60ft), but is usually more associated with shallow reefs where it is food for several species of nudibranch and flat worm. A common species, it occurs throughout the entire region.

2 **FLAT GREEN FEATHER ALGA**

Caulerpa mexicana

Found in sheltered, shallow waters behind protective barrier reefs and in lagoons, this species grows in flat leaves that develop from thin, round runners that are usually just below the surface. The leaves have many pinnate branchlets closely grouped together. It is found in most areas of the region.

3 **OVAL-BLADE ALGA**

Caulerpa prolifera

Often found amongst seagrass, this alga has wide branching blades protruding perpendicularly into the water. The blades reach over 15cm (6in) long and can be as broad as 13cm (5cm). This species is also very common in the Mediterranean and is one of the species favoured by marine aquarists as it is non-toxic.

Red Algae

Taxon: Rhodophyta

4 **LAVENDER CRUST ALGA**

Hydrolithon farinosum

There are a few almost identical species which are known to encrust the Sea Pearls, but this is the most common. It also grows on other algae and seagrass and is common throughout the entire region.

5 **Y-TWIG ALGA**

Amphiroa rigida

Fairly common through the entire range, this alga forms tangled clumps of branching, tubular stalks, usually growing only to around 15cm (6in). It is most often found in shallow water in seagrass beds and shallower coral reefs in a fairly sheltered position.

6 **TUBULAR THICKET ALGA**

Galaxaura rugosa

This species is quite easily identified, due to the tubular shape of the individual leaves. It can grow in clumps up to 15cm (6in). It inhabits protected sandy areas inside the reef and like most algae it prefers shallower waters of less than 12m (39ft). It is quite common in the region.

7 **CRUSTOSE CORALLINE ALGA**

Rhodophyta spp.

There are many different species and forms of this genus of alga, which vary so much that it is nigh-on impossible to identify to a species without laboratory examination. Suffice to say that this highly encrusting alga can be found, in one form or another, in all regions.

1
2
3
4
5
6
7
7

GLOSSARY

Aggregation: Individuals gathered together in a loose group. Can apply to fish or to attached (sessile) animals.

Anal fin: Single fin running along the underside of a fish, from the anal opening towards the tail.

Antennae: Projecting feelers, which have a sensory function.

Appendage: Parts of an organism that project from the body and have a specific function.

Basal plate: A hard plate at the base of some corals from which individual stalks or branches grow.

Benthic/benthos: Of the seabed, and animals and plants that live there.

Carapace: The hard upper or outer shell of a turtle or crustacean.

Caudal fin: A fish's tail.

Cerata: The fleshy extensions or tentacles on the back of a nudibranch, which often store toxins from its prey to help in its own defence.

Chelipeds: A crustacean's claws.

Circumtropical: Found in all tropical oceans of the world.

Cleaning station: A place on the reef occupied by cleaner fish or shrimp, which remove parasites and debris from larger species.

Colony/colonial: A group of animals living together, or an apparently single organism that consists of numerous similar or identical interdependent 'units' (for example, polyps) that are linked together to form the complete organism.

Commensal: Relationship in which two species live in close physical association, with one sometimes dependent on the other for some function or other.

Coralline: Resembles the material from calcareous coral skeletons.

Detritus: Particulate matter of algal and animal origin, which can become attached to the spines of various crustaceans and fish.

Dorsal: The upperside of an organism – see also Ventral.

Dorsal fin: The fin on the top of the back in fish and cetaceans. Species may have one, two or three dorsal fins.

Drop-off: The point at which a submarine plateau gives way to a deeper wall.

Ectoparasite: Animal that lives on the outside of another, and is dependent on it for its nutrition (for example, it attaches to the skin and feeds on blood).

Herbivorous: Animals that feed on plants, such as algae and seagrasses.

Hermaphrodite: Individual that has both ovaries and testes. These may be functional at the same time, or the animal may function first as one sex and then the other.

Invertebrates: Animals without backbones.

Mangroves: Trees that grow in shallow seawater, creating habitat around their roots that offers shelter to many animals.

Mantle: The fold of skin which extends from the body of a mollusc such as a cowrie, and may completely cover the outer shell.

Medusa: The free-swimming stage of jellyfish and some other cnidarians.

Motile: Able to move freely (the opposite of sessile).

Nematocyst: The harpoon-like, stinging spike which certain invertebrates can fire, in defence or to stun prey for food.

Operculum: The hard bony plate which acts as a protective cover to the gills of a bony fish; also the protective cover over a mollusc's opening in its shell.

Pantropical: Found between the Tropics of Cancer and Capricorn around the world.

Paxillae: Small umbrella-shaped structures on the upper surface of sea urchins and starfish, which form an external protective skin.

Pectoral: As in pectoral fin, found on the sides of a fish, usually behind the gills – analagous to the front limbs of an amphibian or other tetrapod.

Pelagic: Living and swimming in the water column, rather than close to or on the seabed.

Pelvic fins: Pair of fins located on the underside of a fish, usually behind the pectoral fins.

Plankton: Plants (phytoplankton) or animals (zooplankton) that float in the water column and are moved by ocean currents. Mainly microscopic, but includes larger organisms, such as some jellyfish.

Polyp: A six- or eight-armed, anemone-shaped invertebrate organism, typically sessile (used to describe representatives of several unrelated groups, including the reef-building coral polyps).

Predator: Animal that hunts and feeds on other animals.

Radula: In all molluscs except bivalves, a toothed structure that is used to draw in and break down food.

Reef: A stable ridge of coral underwater.

Rhinophores: The sensory appendages on the head of nudibranchs.

Scutes: Horny plates that form the shell of turtles.

Sessile: Fixed to a spot (the opposite of motile).

Spicules: Calcium spikes found in sponges and soft corals.

Symbiotic: Relationship in which two (or more) species live together and benefit each other in one or more ways.

Territory: A living, feeding or breeding area that is actively defended from intruders by the inhabitant(s).

Ventral: The underside of an organism (see also Dorsal).

ABOUT THE AUTHOR

Lawson Wood is from Eyemouth in Scotland and has been scuba diving since 1965; he now has more than 16,000 dives logged in all of the world's oceans. He is the author and co-author of 50 books. This is his third book for Bloomsbury, the others being *Marine Life of the Mediterranean* and *Marine Life of the North Sea and English Channel*. He made photographic history by becoming a Fellow of the Royal Photographic Society and the British Institute of Professional Photographers solely for underwater photography, and he is a Fellow of the Royal Geographical Society. Lawson is the founder of the first marine reserve in Scotland, and a co-founder of the Marine Conservation Society and the Berwickshire Marine Reserve.

ACKNOWLEDGEMENTS

Firstly, I must pay tribute to my wife Lesley who supports me in all my endeavours, even when it means that I have to be away from home researching yet another book project. Thank you, I really could not do this without you.

Over the years collecting the enclosed photographs I have dived with many different groups, dive boats and individuals and it is impossible to thank everyone individually, so I give you a collective 'Many Thanks'. There are a few who stand out in the crowd. The tourism boards and PR companies for: the Bahamas; Bermuda; ABC's Cayman Islands; Mexico; Florida; Trinidad and Tobago; St. Lucia; and both sets of Virgin Islands. Individual companies and businesses who have hosted us include: The Aggressor Fleet International; Plantation Resort on Crystal River, Florida; Pura Vida Divers at Blue Heron Bridge, Florida; Florida University; Trimarine in the Virgin Islands; UNEXSO; Stuart Cove's Dive Bahamas; Steve Broadbelt at Ocean Frontiers, Cayman Islands; Nancy and Jay Easterbrook on Grand Cayman; Little Cayman Beach Resort; Cayman Brac Beach Resort and their various and numerous dive staff; Martin and Liesel van der Touw and the Brac Scuba Shack; Scuba St. Lucia and Anse Chastanet; Paul Humann and Ned DeLoach; Helmut Debelius; Nick Davies and Jennie Aguilar of Belize Travel Services; Perry Bodden from the Belize Tourism Board; Liz Parkinson; Sean Robinson and Tobago Dive Experience; Captain Don's Habitat on Bonaire, Curaçao and Saba; and Alan Marquardt in Bermuda. Special thanks to Professor Mary K. Wicksten, Department of Biology, Texas A&M University; Christopher Pomory at the University of West Florida; Anja Schulze of TAMU-Galveston; Lisa Collins and Mateusz Bednarczuk of Capture Cayman; Fimscan UK Ltd; Cayman Airways; British Airways; and Bahamasair.

PICTURE CREDITS

10 Jose Mario Espinoza/Shutterstock; **31** (3) John Bantin; **33** (5) Luke Atkinson; **51** (5) Peter Douglas Clark/Shutterstock; **65** (3, 4) Lisa Collins; **69** (7) Lisa Collins; **71** (1) Lisa Collins, (2) Alex Mustard/Nature Picture Library, (4 inset) Mateusz Bednarczuk; **75** (7) Lisa Collins; **77** (5) Lisa Collins; **81** (2) Anita Floyd; **85** (1) Lisa Collins; **91** (2) Karel Bartik/Shutterstock; **107** (1) Peter Leahy/Shutterstock, (5) Lisa Collins; **109** (2) Joe Quinn/Shutterstock, (5) Claudio Contreras/Nature Picture Library; **111** (4) Jesus Cobaleda/Shutterstock; **113** (1) Lisa Collins; **117** (2) Charles Hood, (3) Shane Wasik; **119** (3) John Bantin; **125** (4) Lisa Collins; **139** (8) Lisa Collins; **147** (4) Lisa Collins; **149** (1) Lisa Collins; **151** (2) Lisa Collins; **153** (7) Lisa Collins; **157** (2) Lisa Collins; **161** (2, 3, 5, 7) Lisa Collins; **183** (6) Lisa Collins; **185** (2) Lisa Collins; **189** (5, 6, 7, 8) Lisa Collins; **191** (4, 6) Lisa Collins; **193** (4, 5, 6) Lisa Collins, (8) Malcom Nimmo; **195** (4, 5, 6) Lisa Collins; **197** (1, 2, 3, 4, 6, 7) Lisa Collins; **207** (8) Lisa Collins; **213** (3) Lisa Collins; **217** (6, 7) Lisa Collins.

With thanks to Julian Baker for redrawing 24 illustrations on pages 25–29.

INDEX